Mr Nobody

Almost Famous, but not quite

– DESMOND MCGRATH –

Printed and bound in England by www.printondemand-worldwide.com

Old age is earned.
The alternative is far less popular.

Desmond McGrath
T.A.R.P.

http://www.fast-print.net/bookshop

Mr Nobody

Copyright © Desmond McGrath 2018

A catalogue record for this book is available from the British Library

ISBN 978-178456-57-32

First published 2018 by
FASTPRINT PUBLISHING
Peterborough, England.

DESMOND McGRATH

MY AUTOBIOGRAPHY

MR NOBODY

DEDICATION

To Kimberley my first born granddaughter, whose idea it was to write the book. Also Ella my eleven year old granddaughter, who always shows a interest in everything I do, especially my writing. When I told her about the book she said, "Am I in it Grandad?"

And also the rest of the grandchildren in order of birth, Jack, Tara, Chloe, Jensen and Livy.

PART ONE
THE BEGINNING

Prologue

My dad was on the run. I wasn't yet born. Spike Island was a British military base off the coast of Ireland. I don't know what Dad was doing there but must consider it was illegal. It was open water and his accomplice, Seamus, and he were fleeing for their lives. They were rowing furiously and were frightened and exhausted.

Seamus wanted to surrender. Dad didn't. There would be serious consequences if they were caught. An argument ensued and it was turning violent. Rifle barrels flashed and flamed with loud cracks. Bullets from Spike Island were splashing like jumping fish all around the boat. A splinter flew from the boat as a round struck into it. My dad was cut and bleeding, as was Seamus.

"We have to give up, Paddy," Seamus cried.

"Like hell we will," screamed my dad, standing up in the boat. "If you don't row this boat, I'll cave your bleedin head in."

A bullet crashed into his oar, almost knocking it out of his grip. "Row the boat," he screamed, sitting back down in his seat.

They both rowed furiously to get out of range. The bullets began to splash farther behind them. They were out of range. In the pitch black of a moonless night, Spike Island was looking smaller by the minute.

The whole operation had been a total balls-up. They had achieved nothing but a bullet-riddled boat.

"We'll have to sink the boat," said my dad to a shivering Seamus. "But it's the only boat I have, Paddy," protested Seamus.

"Is it worth going to jail for, Seamus?" asked my dad, "Because it isn't to me."

"I'm not a strong swimmer, Paddy. I'm scared," wept Seamus.

"Me neither, mate," confessed my dad. "That's why we've got life-jackets. Now get one on. We'll get as close to the shore as we can then scuttle her. It's the only way. If they don't find the boat, they might not find us."

"To be sure, I see that, Paddy, but me boat!" Seamus said, with quiet resignation. "You can always steal another one, Seamus," grinned my dad.

The breakers were crashing across the shingle beach like a foaming bubble-bath.

"We're as close as we can get," my dad said grimly. "The water's deep here. Put the axe through the bottom and we'll just sit here until the boat disappears from under us. Keep calm and good luck."

"To be sure, Paddy," choked Seamus. "'Tis luck we'll be needin' tonight, all right."

The boat slid down and away from under them. They were up to their shoulders in choppy, freezing, black water.

"Swim, Seamus," cried out my dad. "The closer you get to shore the easier it gets. The waves should carry us in."

They swam for their lives, tiring with every stroke. The crashing of the breakers was getting louder and louder. They were inching mercifully closer to safety. Or capture. The swell beneath was lifting them up and forward. They were almost there. A thunderous wave broke over them, thrusting them closer to safety. Shingle scraped and clawed against their feet. They found grip.

Tired arms clawed at the tiny, shifting stones. One final wave threw them half onto the beach. The surf receded, to leave them high and dry.

They rolled onto their backs, gasping for breath. They were on a dark, deserted, cold, lonely, windy beach. They sat up finally and threw aside their life-jackets.

"'Tis out of here we better be getting, Paddy," gasped Seamus. "They probably radioed ahead. They might be looking for us."

They stood up, soaked and frozen, and shook hands firmly in the dark. "We'll split up," gasped my dad. "It'll be safer alone."

"Good luck, Paddy," said Seamus. "God save Ireland."

"Good luck to you too, Seamus," returned my dad. Then turning away, to himself, "God save me. Fuck Ireland, I'm away from it."

He told me this story when he was an old man. He never told me who Seamus was and what they were doing on Spike Island.

That remains a mystery to this day.

One

He needed to get away from Ireland. Fast. He gathered up my mother, his girlfriend at the time, and together they fled to Jersey in the Channel Islands. They rented a small, back-street cottage in the centre of St Helier. 4 Brook Cottages, Ann Street.

There were no jobs for butchers but there was a thriving building industry. My dad managed to talk himself into a job as a foreman painter, for Cranwell Housing.

Mom fell pregnant with me and eventually I was born in hospital in St Helier. In the bed next to her was a girl her own age who had also given birth, to a girl, the day before.

June Hansford was the woman's name. Her baby was Janine. Mom and June became close friends - a bond that would extend even after their lifetime through the friendships of their children.

Seventy years later, the friendship is still as strong.

June's husband, Roy, was to become my dad's best friend and drinking companion. Janine and I would always be close. My brothers and her brothers and sisters would also become friends. In fact, more than that.

Family.

To me they will always be my cousins. June and Roy, I always called Aunty and Uncle.

That's how it felt.

My first memory of life is a very young one. I was still in nappies and barely walking. I am stumbling around in a barn with an old black car in it and rabbits running around my feet. Mom tells me I must have been about eighteen months and the old barn was at 4 Brook Cottages.

The next thing I clearly remember is standing up in a hospital cot in Kent, screaming and terrified because my mom was leaving me there. I had suffered a perforated eardrum. It was an experience I was to take advantage of in later life as an excuse for my selective hearing.

On my release from hospital, we stayed a short while at my dad's brother, Gerry's, in Sidcup, before moving on to Birmingham.

We were living in two rooms with my mom's brother, Eddy, and his wife, Mary, in a house they owned in a lovely area, Small Heath. It had long, tree-lined streets with a shop on nearly every corner. There were very few cars, and the milk was delivered on a horse-drawn cart.

At four years old I used to love chasing down the road, petting the horse. However, at that age I could not understand why people were following the horse with a bucket and shovel collecting up the steaming poo. I later learned it was to put on their rhubarb patches. Apparently, everybody had a rhubarb patch at the bottom of their garden.

Perhaps that's why I don't like rhubarb.

Our address was 100, Somerville Road. It was a very long road. Further down it lived my grandad and granma, Pop and Mud. That's all they were ever known as, Pop and Mud. I know everybody says the same thing about their gran and grandad but, to me, they really were the best in the world. More about that later.

Two

Mom and Dad were grafters. Mom scrubbed floors in the hospital. She saved every penny. She wanted a house of her own.

My dad was a butcher and was managing a shop for Mrs Barker in Aston, Birmingham. He was ambitious, very ambitious, and was also saving his money. He wanted his own shop. He was to get it. And not just one, but many.

The house came first. 63 St Oswalds Road, Small Heath, just off Somerville Road and down the road from Ed and Mary and Mud and Pop. It was a large terrace house with a huge, long back garden.

I was five by then and ready to start school. Three doors away lived a family, the Horgans. They had a son, Kevin, same age as me. We became instant friends. And still remain friends to this day although, as is inevitable with time, lives and circumstances change and, with a widening circle of children and grandchildren, you lose touch a little. But we both know that we are there for each other if needed. As proved to be the case a few years ago when my mother died - Kev was there.

We started school together at the Holy Family School in Small Heath. They were to be four happy years.

However, I was a sickly child and through the years lost much time and many lessons. If there was anything going around, I always seemed to catch it. In the last year, in a playground accident, I broke my wrist and was in plaster for three weeks.

It was time to take the Eleven Plus. The exam of the times. It decided what school you went to next, a grammar school or a secondary modern. I failed the exam miserably and was sentenced to a secondary modern.

The shame of it. However, I was not alone; at least my best friend, Kev, was coming with me. Six weeks later we were starting at The Rosary, a Catholic school run by Marist Brothers. It had a reputation as a tough school. I was soon to learn that there were two ways to survive. You could be tough and fight. Or you could be popular and make them laugh!

I chose tough!

"What?!" I hear you shout.

"Hey, just kiddin'," I say.

See what I mean? Make them laugh.

Just when I thought it couldn't get any worse, it did. First day, everyone in the school is lined up in the playground to see what class they will be in. And what teacher they will get.

There were four years. Each had three classes. A, B and C. I'll list them.

A   Not so dum.

B   Dum.

C   Dummest of the dum.

You've guessed it. Yes. I know. I was in the C Class. Dummest of the Dum.

However, after leaving the Holy Family and taking up my position at the bottom of the Rosary, there were the summer holidays.

And that meant sun, sea and sandy beaches. Jersey. The annual trip to stay with Aunty June and Uncle Roy. They were a large family in a three-bed semi on a council estate, but there was always room for us. My brother Phil was only four. I was eleven. Mom spent most of the day on the beach with Phil. I spent the days with the kids. Royston was the older, "Bashy" by nickname. Janine, who was born in the next bed. She was always called Tuppy. Then there were Bridy, Dennis, John and Joel. Susie, the oldest girl, did not live at home. She was married, I think.

My day started early, which was to become a feature of my life. At six in the morning, me and Bashy had to go to the local bakery and collect a couple of big brown bags of steaming-hot bread rolls.

The smell. I'll never forget it. It was amazing. I can still smell it now. It is one of the most vivid memories of my childhood. I can't remember what we had with it, but we were set up for the day.

And what days.

I have no memory of rain. Just long, endless, sunny days. The beach was five minutes' walk from the house. When the tide was in we dived into the surf. When the tide was out we explored the huge rocks and pools.

Other days we went fishing. They started by digging for bait. The best place was the mud, when the tide went out at Elizabeth Castle. Up to our knees in thick, filthy mud we dug up big, fat lugworms. They were long, fat, hairy and disgusting.

But with a bucketful, it was off to the docks at St Helier. We weren't supposed to be there, but Uncle Roy was a big noise around there, so we were OK. We fished between the moored-up merchant ships and passed the time making friends with all sorts of navy crews. Every nationality you can think of. It was great. Whatever language they spoke, you always managed to communicate. Some would let us go on board and fed us all sorts of strange fancy nibbles.

The only fish we ever caught were small. Flobbers, they called them. They fed on the waste from the ships. Bags full, we set off home to Aunty June's.

She washed and cleaned the flobbers and fried them crisply and, with another bag of hot, fresh rolls, dinner was served.

Other days we spent exploring the German bunkers. It was only about fourteen or fifteen years since the Germans had left. You could get a green bus anywhere and days quickly passed, exploring and searching for souvenirs. In those days you could still dig things up. A mug, a plate, sometimes even a helmet.

The six weeks flew and soon it was teary farewells and back to Birmingham.

To school.

To the Rosary.

And, to be fair, it wasn't as bad as I'd feared. There were the usual bullies and tough guys. Most of those were in my class. The C Class. However, my teacher in that class soon decided I was in the wrong place. At the end of the first term, I was put in the B Class. The second term was Christmas to Easter. At the end of that term it was decided I should start Easter to summer term in the A Class.

The year-end summer exams came and I was second in class. The next three years I was first in class each year and left the Rosary as School Captain.

So much for the Eleven Plus.

• • • • • • • • •

My apprenticeship as a butcher had begun early. Eight, to be exact. Holding my dad's hand, I would go to the market in Bradford Street to choose the carcases he wanted to buy. He and all the salesmen would point out what to look for in a slaughtered beast. Over the years, I was to learn that you made most of your money at the market. The difference between a good beast and a bad beast was the difference between profit and loss. And if you couldn't tell the difference, then the salesman wasn't going to show you.

I was not a little boy any more. It was dog eat dog. The negotiations over price were tough and ruthless. You never paid what they wanted unless you really needed it. And God help you if they sensed you needed it. A poker face was essential and you had to know when to turn your back and walk away. And how fast and far to go, heart in mouth, waiting for a shout, "Des".

You stopped, looked at the beasts on the next stall then turned. "What?" When they called you back, you returned reluctantly. A handshake, and the deal was done.

But my dad always warned me, "If they don't call you, never go back."

Over the years until I left school, I would go to the shop after lessons and help clean down. Saturdays were spent all day learning the cuts, boning-out and delivering orders. I got paid fairly and collected tips from the deliveries.

I opened an account at my dad's bank and saved every penny I could. By the time I left school I had a considerable stash.

It was taken for granted that I would join my dad in the shop full-time when I finished my education. I was good with that and was paid a fair wage.

It was long hours and hard work. Up at five in the morning, off to market and back to the shop to put on the window. A good display sold the goods the Old Man always told me. Pile it high!

When all the prep was done we had a big fry-up for breakfast and the Old Man set off back to the market. It wasn't far from the shop. At ten in the morning the market was almost all cleared up.

He was what they called then a clear-up merchant. He used to stroll from stall to stall asking if they had anything they wanted to get rid of. The Old Man would offer them a silly price and clear it all up. That was the thing. You couldn't pick and choose. You had to take the lot. Whatever it was, we always managed to sell it.

And that was where the money was. As the Old Man was fond of saying, "Anyone can sell fillet steak - give me the man who can sell pigs' heads."

My agenda at the time was to save every penny I could and, when I was eighteen, join the Merchant Navy. I planned to see the world.

And not swabbing decks!

I was going to be an officer. A Radio Operator.

"Why?" I hear you ask.

Because I had found out that was the easiest way to become an officer. And I wanted that white cap and uniform. I dreamed of myself drinking beer in far-off ports and harbour bars, beer in one hand and my arm wrapped around some beautiful foreign girl with brown skin and long, shiny black hair, and little else.

Obviously, nothing goes exactly to plan but, all in all, apart from the Navy and the uniform, things didn't work out too bad.

So, I'd left school in July and was preparing myself for life's great adventures. What they would be, God only knew.

Only time would tell. Was luck on my side? Or do you make it yourself? I don't know. But I'm not complaining.

But before all this was to begin, there was one last scene to play out with the Rosary School.

Three

Switzerland

The past-pupils' club had organised a trip to Switzerland. Four of our class had saved up all year for it and now it was here. Myself, my best friend Kevin Horgan, John Maloney and Eugene Reynolds arrived early one Friday evening outside the school to board the coach. We were the only pupils from our year. The rest of the group were past pupils of various ages.

Brother Bede and another teacher were the school representatives. Our moms and dads were there to see us off and give us all the usual warnings.

Be careful, don't do this and don't do that. You know, all the usual concerned-parent stuff.

We reassured them, as you do, and took our seats on the back row of the coach.

We were all excited. It was our first time abroad. It was 1963. The Swinging Sixties. Part of the group was a band. Marco and His Men. They had all their gear and were hoping for a few gigs along the way. All the rest of the passengers were much older than us. Some were even couples. We never were to mix much with them as it turned out. But, after a head count we were off on our way. At fifteen, to us it was the start of a great adventure.

The first of many in our lives.

After a few hours of excited chatter, it went dark and we settled down in our seats to sleep and dream. Girls, girls, girls.

We arrived in Weymouth in the morning around five. It was good to stretch our legs for a bit and find a toilet. We had about an hour before we had to be back on the coach to board the ferry. To be honest, there wasn't much to do and the only thing I really remember was pinching bottles of milk off somebody's doorstep.

A few hours later we docked.

Belgium. Brother Bede took care of everything and soon we were back on the road. Mile after mile we sped through the beautiful towns and villages. We stopped at prearranged places for food and the toilet and eventually arrived at our first stop for the night in Reims.

The hotel was basic and the food was fantastic. The four of us had a room together and it wasn't long before we set off to explore the town. I don't know what the drink laws are in France, but we had no trouble getting beer.

Dressed in our round-neck foam Beatles jackets, drainpipe trousers and winkle-picker shoes, we cut quite a dash. The locals loved us and were so friendly. We found a bar that was full of young people. Not aware of the local customs, we tried to tread softly and go with the flow. We could barely speak a word of French, but the locals seemed to speak enough English for us to communicate. One of the local customs it seemed was that when a girl came in the bar she would kiss every boy in the room. Including us! Great custom. And, of course, when they found out that we were a travelling pop band from Birmingham on our way to a gig in Switzerland, the kissing seemed to intensify.

"THE MODS", our band, had just signed a record deal and were releasing our first single, Diabolical Love Affair, the following week, and Switzerland was the first gig of our continental tour.

Sadly, though, we had a management curfew and we had to get back to our hotel. So, with much hand-shaking and kissing, and the odd touch-up, we left our excited fans and sourced a couple of bottles of red wine for the room.

Well-pissed and laughing outrageously loud, we were horrified to hear a knock on the door. A very angry Brother Bede, by the sound of it. In a bit of a panic, we hurriedly opened our window and stashed the wine outside on the ledge. Opening the door, we were confronted by a furious, red-faced Bede.

We obviously denied everything and, after a search of the room and a severe bollocking, he left. Leaving it a few minutes to see that it was safe, we opened the window to retrieve our booze only to find that it was gone. It had fallen from the narrow ledge and smashed into the alley four storeys below.

There wasn't much left, anyway.

Next morning it was a hearty continental breakfast and back on the road for the final leg of the journey to Interlaken. The scenery just made you gasp. When the coach arrived at our hotel, the setting left us speechless. Set in to a mountain and overlooking a huge lake, too, for fifteen-year-old lads it was something from a story book.

To us it was a hotel. In all probability it was a youth hostel. We were allocated a large dormitory for eight. It was then that we were introduced to Marco and His Men. They were four boys older than us, seventeen or eighteen years old. We quickly became friends and didn't object when they asked if they could set up all their gear. There were a drum kit, speakers and mics and guitars. In the day, when it was quiet, they were allowed to practise. They weren't bad, and I can't recall their names, but it was all great fun, with us joining in from time to time.

However, there's only so much you could take and we preferred to go off exploring. With towels and trunks, we headed to the lake. There was an area of concrete on three sides that was open to the lake on one side, and a small diving-board. The water was as clear as crystal and the fish could swim into the pool area. It was alive with fish. It was late July and the sun shone from a clear blue sky. It was a lovely temperature, a constant 70°F or so. It looked like a tropical lagoon. We couldn't wait to get in. Myself, Kevin, John and Eugene lined up on the side to dive and cause havoc.

Splash.

We hit the water and screamed. Not joy. Horror.

It was absolutely freezing.

It was like diving into the Arctic Ocean. Nobody wanted to be the first out. I gave it five minutes. Not being the most heroic of people I shouted, "Stuff this," and scrambled out, rapidly followed by the other three wimps. Shivering with hypothermia, or so I called it, we never went swimming again.

Cossies back in the case.

So, it looked like we needed to find some other activity. Next door to the hostel was a Swiss chalet-type bar. All pinewood character decor. We ordered four beers and were delighted to find them served in big glass jugs with large handles you could get your whole hand into. As with France, we didn't know what the local licensing laws were, but we were all quite big for our age and looked older and acted older than the fifteen years old we actually were. Anyway, we got served.

Most days were set aside for day trips in the coach, so we saw a good deal of the country. One memorable trip was to Lucerne, where Kevin and I hired a rowing-boat to splash around the lake. On returning to the jetty, Kevin swept his oar across and took the legs out from under the young kid who was waiting to tie it up. Not at all impressed with our boatmanship, he shouted out and proceeded to give us in Swiss what I can only imagine was a mouthful of abuse. Kevin, not impressed with his manners, hopped on to the jetty and took a swing at him and, before a full-blown fight began, I jumped out and calmed things down before we had an English-Swiss international incident on our hands.

Kevin always had a short fuse and me, not wanting to get my head punched in, began a lifelong love of diplomacy.

Discretion and valour and all that.

So, on the last Friday of the holiday, Brother Bede asked the group who would like to go mountain trekking.

About twenty of us.

Behind the hostel was a very large mountain. There was a trail that took long zig-zags up the mountain from left to right. It was a safe trail that looked like it would take forever to get to the top.

Brother Bede led the procession, with everybody following in twos. Kevin and I were the last. I liked it that way. I didn't like being hemmed in by other people. Still don't. Looking up the mountain, you could see the route of the trail. Long sweeps snaking left to right.

Me and Kev seemed to be trailing behind a bit, so I suggested to Kev that it would be a lot quicker if we went straight up and met the others further up. It wasn't steep. It was just grass and small rocks. Bit like the Malvern Hills, it seemed to me.

"OK," said Kev.

And that was that. Pretty much on all fours we started scrambling our way up. It was lovely and sunny and soon we were loving it. We'd long lost sight of the others but knew we would come across them on the path somewhere above us.

We soon, though, realised that we hadn't actually crossed the trail for a while on our way up. But, with my navigational skills and sense of direction, it stood to sense that up was up and sooner or later we had to cross their trail, or at least come across one. Then it was only logical that down was down.

Trouble was we never came across another trail. We realised that we weren't going to get to the top. We then realized that we'd been gone a good three hours. We were hungry and we were lost. And there was no trail down.

Suddenly, the terrain seemed to be a lot wilder, we were deeper into the trees. We came across a stream a few inches deep with a smooth, rocky bottom. We cupped water from it in our hands. It was cool, pure and sweet. We sat down beside it and took stock of the situation.

"Well," I said to Kev, "I reckon if we follow this stream down we can't go wrong."

Kevin agreed. However, we found out it was a lot harder to get down than it was to get up. It seemed an awful lot steeper to me. It was. You couldn't walk down. We had to slide on our backsides. The grass was a couple of feet long, so we used to hold a handful between our legs and use it as a brake.

We hadn't really noticed how quiet it was up there until we heard what seemed like loud crashing in the trees. We pulled on the brakes and sat listening. Kevin was on one side of the stream and I was on the other. It was about six feet wide. As we looked and listened, out of the trees strolled a huge, magnificent, reddish-brown deer with a massive pair of antlers.

It stopped at the edge of the stream. It tossed its head a couple of times, snorted and looked straight at us.

SHIT.

Kev looked at me. I looked at Kev. The deer stamped its feet a couple of times in the rocky stream.

Double shit.

With another toss of his head and a few snorts he stared straight at us. Then, calm as you like, he lowered his long neck to the water and started to drink. Done, he shook his head a few times and trotted across the stream and into the trees.

A wonderful brush with nature that is still as vivid today as it was then. Chattering excitedly to each other across the stream, we continued our way down and eventually emerged from the trees into the sunshine again. It was late afternoon by now and, some way off to our left, at the bottom of the mountain, we could see the hostel and the zig-zag trail we started on.

We slid down to it and followed it back home.

To the biggest bollocking we had ever had in our lives.

Brother Bede went ballistic. In language unbecoming a man of the cloth he screamed, "Where the bloody hell have you been? We were just about to send out rescue teams."

We survived the bollocking and, as it was our last night, got well and truly pissed and packed to go home. Marco and His Men did a gig in the room and nearly got us all thrown out. They decided not to come home with us and stayed to make their fortune.

We had one more kissy-kissy night in Reims, flushed with the success of our continental tour, and arrived back in Birmingham to begin our lives in the real world.

And made a page in the past pupils' magazine.

There was a full page describing our adventure on the mountain and telling how, with our ingenuity and survival skills, we had managed to save ourselves just as a search party was organised to go and find us. There was a photo of us with Brother Bede beaming to the camera.

Happy days.

Four

Work - The Real World

It was 1963. The Beatles ruled the charts. I loved them. Their next single was eagerly awaited by me and the whole country. The Mersey Sound inspired a whole generation. New bands were emerging all the time. The Searchers, Hollies, Herman, Mersey Beats, Billy J Kramer. Too many more to mention. Everyone had their favourites. My early heroes, Adam Faith and Billy Fury, were being pushed aside.

I was working hard and earning good money. And stashing it for my time in the Merchant Navy. A year flew by like lightning and, a few weeks after my sixteenth birthday, my dad bought his second shop. A run-down business in Small Heath in a very promising area inhabited mainly by well-off English natives and ambitious Irish immigrants who loved their meat. It was a densely-populated area of road after road of terraced houses.

A goldmine for someone like my dad. An Irish butcher in an Irish stronghold. All he needed was the right man to run it. But that was the big problem. They were thin on the ground. That was the only flaw I could find in his plan.

But nothing ever stood in the Old Man's way. He had the very man in mind. He didn't say who it was but, knowing the Old Man, I was sure he had it covered.

And, of course he had.

ME.

But hang on a minute. What did I know about managing a shop? I was a good butcher, yes. But the responsibility at sixteen. I wasn't ready for that.

"No problem," said the Old Man, brushing aside my concern. "Just give me twenty-five quid a week and pay all the expenses and the rest is your own."

Well, twenty-five quid was a lot of bloody money in those days. I couldn't drive and didn't have a van. No worries there, he'd drop all the meat off and all I had to do was sell it.

I was living at home at the time. The plan was that I go to the market with him and get the meat; he would then drop me off back at the shop and go to his shop and open up. At the end of the day, he would come and pick me up and drop me back home.

I did a week in the shop with the old guy that had sold it, to get to know the existing customers. The old guy was retiring and wanted to visit his son in Australia. He'd been there most of his life and just wanted to get out quickly.

Always willing to help anyone out the Old Man paid him, in cash, a fraction of what it was worth. The potential was enormous and nobody in the trade could understand how he had managed to buy it without it ever coming on the market.

The night before my opening day, I was physically sick with nerves. I couldn't sleep. Five in the morning we were up and off to the market. In those days nothing was prepared. You bought whole sides of beef, whole lambs and pigs, bags of offal, even a cow's head.

It all had to be cut and boned. I was working flat out all day long and serving customers in-between. The first week I almost doubled the trade.

Saturday. That was the big day. I was shitting myself by then. I just couldn't face it alone. That was when my mom's brother, Uncle Bill, stepped in. He wasn't a butcher. He worked at the Triumph factory in Coventry. Trouble was he worked Friday nights, so he had to come straight from the factory to the shop and work all day. The extra pair of hands and the moral support saw me through.

We were more like mates than uncle and nephew, playing golf every Sunday and having a few beers together. Without him I never would have survived. It took over six months for the nerves to settle. Friday nights I just couldn't sleep, worrying about Saturday.

Bill learned quickly and was soon able to cut and bone. His only failing was his familiarity with the customers. He seemed far more interested in chatting up the women than cutting up meat.

As business thrived I found myself awash with dosh. It was my seventeenth birthday and I had an intensive course of driving lessons booked. Three a week for six weeks. Eighteen lessons. My test was booked. Six weeks after, I passed. The Old Man had a van and a new Ford Cortina. He was always flash. He let me drive the van, but not the

Cortina.

It was the summer of 1966 and I wanted my own car. I knew what I wanted. A Triumph Spitfire. There was a showroom close to the shop. I paid them a visit on half-day closing and, a week later, took delivery of my brand-new Spitfire. Powder blue with wire wheels. They were extra, but I had to have them. Six hundred and sixty-six pounds it cost me, and a staggering eighty pounds to insure it.

Boy, was I proud.

I always had my suits made to measure in those days. Burtons in New Street, Birmingham. I treated myself to a beautiful, dark-brown wool one with a velvet collar and pocket-top.

The dogs or what?

Fridays were my big nights out. Best friend Kevin and me would go boozing and cruising.

Our favourite place was the *Carlton Club* in Erdington. It was a fantastic venue. Live bands, booze and girls.

Sometimes they had big acts. The most memorable were Billy Fury and the Four Tops. Fury was amazing and the Tops wore no socks. That fascinated me. Why couldn't they afford a pair of socks?

Christmas was the biggest nightmare. I hate turkeys. The only good thing about Christmas was counting the money.

It was 1967 and I hadn't had a holiday in three years. It was time to put that right. Me, Kevin and two of our old school friends, Danny Peters and Brian Brosnan, started planning a lads' holiday. Two weeks in Jersey.

Hello Aunty June.

Five

Jersey 1967

I'd always wanted to write from as early as I could hold a pen. I wrote my first book when I was fourteen and fifteen. Yes, it took that long because I typed the whole thing with two fingers. Having finished it, I didn't know what to do with it. Obviously, at that age, nobody took me seriously and I couldn't get anyone to even read it. It's still in the loft somewhere. I must get it down some day and take a fresh look at it.

Writing books is hard work. Ask anyone who's written one. So, I turned my hand to song-writing. Couldn't have picked a better time. And it was a whole lot easier. One page of lyrics. Job done.

And so it came to pass that, early in the year, I wrote a song, *"Diabolical Love Affair"*. I sent it to a guy called Nick Sandys, who was advertising for songs in the New Musical Express.

On really posh letter-headed notepaper, he praised my song and said he would be interested in working with me. The letter heading was *Northern Songs*. Conveniently, from the end of his reply to the "yours faithfully" bit, there was a big spacing. Everything was typewritten in those days, so I very carefully typed in an extra paragraph.

"It has been arranged for Manfred Mann to record the song when finished."

It looked really great.

Just wait till the girls saw that.

I took the letter to Jersey.

We landed on the island. Pissed as usual. Start as you mean to go on. June met us in a dinky small car and, somehow, we all squeezed in, luggage and all. Exhaust almost scraping the road. She took us home to her house.

June had cleared the kids out of one of their rooms for us. They didn't seem to mind. Dennis, John and Joel were the lads. Bridy and Janine (Tuppy) were the girls. Tuppy, if you remember, was the one in the next bed born the same time as me.

I was very disappointed to note that she was about seven months up the duff. I'd quite fancied her. Not the best start to the holiday.

Our room had a double and a pair of bunk beds. Me and Kev took the double, Danny and Bros, as we called him, took the bunks. Dan up, Bros down.

Danny was the serious one, very cautious in his ways. Bros, "The Professor", was good for a laugh. He was tall and skinny and wore wire-rim glasses. He would have been great in a pub quiz if they had had them at the time. He knew a bit about everything and not enough about anything.

Kev was like me. Always ready for a play-up. Especially with the girls.

We hunted together.

June's husband, Roy, was a fantastic character. In his fifties, he'd been through the years of occupation by the Germans and was a strong trade unionist. Jersey's answer to Red Robbo.

He took no shit. From anyone.

We had a few good piss-ups in the *White Horse,* I can tell you.

It was Saturday night and we wanted a good time. Large. *The West Park Pavilion.* That's the place, Roy told us.

"Bags of crumpet," he howled with laughter. "Bags of crumpet."

That afternoon we hired a Mini Traveller, the one with the wood sides. It was brand-new and, as I was the only one who could drive, I was the chauffeur. I didn't mind, I liked driving. Just being cautious, I asked Roy how the police were with drink-driving. He enlightened me.

"Desi," he always called me Desi, "You've got a big H on the back of the car so, unless you do something really stupid, they'll leave you alone. Now if you do something mildly stupid, June's a traffic warden and should be able to fix it. If you do anything really stupid, then one of my best mates is the Deputy of the Island. If you have an accident, you're screwed."

Armed with this valuable information, we set off for The West Park Pav with three things on our mind. Get pissed, don't have an accident, and crumpet.

The Pav was a huge dance-hall on the front, just outside St Helier. Up on a hill and all lit up, it was like our *Locarno or Tower Ballroom.* If you're old enough to remember. If you're not, ask your mom or dad.

I've never been much for dancing. Still don't like it today. And as for chasing girls, the same rules apply as trying to photograph fish under water. I'll come to that later. Swimming about the place trying to find fish never works. They just swim away from you. The trick is to find a quiet lair and keep still. Eventually curiosity overcomes them and they come to you.

Simples? Yes!

The best lair, I always found, was the corner of the bar. Preferably on a stool. I've always liked a cigar, and the choice of smoke at the time was Manikin. Kev liked one too. Danny and the Prof didn't smoke. It was a big bar, with plenty of room for us to hang.

Right on cue, edging closer, were a couple of real crackers. I told Danny and Prof just to try and give them a little space. Casually, I sipped my beer and acknowledged them. Just a polite, disinterested nod.

"Are you here on holiday?" my one asked. I'd made up my mind which was mine. "Well, yes and no, sort of," I said, as if I were a bit confused.

"What does that mean?" she asked, in a heavy northern accent.

I love Newcastle girls. They're so friendly.

"Well I was born on the island," I told her, truthfully, "But I've been working on the mainland. I'm taking a break from the studios and come home to spend a break with my family."

"The studios?"

"Yes. I'm a songwriter and I've just spent a gruelling three months locked up in a studio packed with musicians, tech men and tossers. I've been helping The Hollies with their new album."

"Get off with yer," she laughed loudly, "That's a new one."

"Excuse me, if you don't mind," I cried indignantly, "But I've just had a chart No1 with The Hollies. Bus Stop. And, if you must know, I'm working at the moment with Manfred Mann."

"Honest to God?"

"Here you are," I said, pulling my letter from my inside pocket. "If you don't believe me, read this."

While her mouth was opening slowly wider, I could see that Kevin was taking up the rear guard with some equally brilliant bullshit of his own, as he closed his arm around his girl and ordered a round of drinks. Girls included. I think he was the manager-cum-sound engineer, or some crap like that.

So, with the midnight hour striking and the girls nicely lubricated, it was time to go home. It was a tight squeeze, six in a Mini Traveller. But I'd got it covered. Kev and me in the front, and the girls behind. Don't ask me their names. I haven't got a clue. And, come to think of it, I don't think I ever asked. Then there was Danny and the Prof. They fitted nicely in the luggage space behind the passenger seats. I had to tell them to shut up moaning, or get out and walk. They were a little stiff as I let them out at June's place.

Actually, I was feeling a bit stiff myself. I found a nice quiet spot along the coast road, where we could stretch out. Kevin was feeling that stiff he decided to take a stroll down the beach with girl B. With a lot more room in the car, I was able to stretch out with girl A and soon felt a lot more relaxed.

A couple of hours later we dropped the girls back to their B&B and returned home to June's, feeling completely shagged-out after a long, hard night. We slept soundly.

Breakfast never changed. Delicious hot rolls.

The two weeks were whizzing past fast. Horror of horrors, the two girls turned up, knocking on June's door looking for the songwriters. I can only think they remembered from when I dropped the lads home. Roy told them exactly where they could find us. NOT. He sent them to the Watersplash down the Five Mile Road. Cruel bastard. It was about as far as you can get from June's.

As this isn't a novel, I'm not going to go into every detail of the two weeks. It was all much of a muchness.

We toured the island by day, drinking, swimming and sunbathing. We toured the island by night, drinking, eating and hunting. Danny and Bros stayed true to their beliefs and remained virgins. By choice or design, I'm not sure. But I have my suspicions.

The lovely, plump Tuppy remained the only disappointment of the holiday. The kids got their room back and, even so, seemed genuinely sorry to see us go. The girls shed a tear and Roy had a lump in his throat as we waved goodbye to the house, and Aunty June drove us back to the airport.

As our flight was announced, June opened the floodgates. And started me off. We thanked her for her amazing hospitality and, after much genuine refusing, she reluctantly accepted the generous wedge we forced her to take. They weren't a rich family and would go hungry themselves to feed you. A great family and great friends. I love them dearly. It would be only a matter of time before June would arrive in Birmingham to stay with my mom. She knew she'd get the same reception.

It was a short flight, but we managed a beer and a new drink we had grown fond of - Bacardi and Coke. The plane landed and we were home.

Back in the shop again, it was all systems go. It was great fun. I loved the crack with the customers. Some, I know, only came in for a chat and the banter. One of those customers was to end up my first wife.

Rita was almost seven years older than me. I was seventeen, she was nearly twenty-four. She lived in Monica Road, a few roads away, with her dad and brothers. She was fun and attractive, and it wasn't long before I asked her out. I was so much older than my age and it was good to have a girlfriend. We hit if off quickly and soon I was seeing her two or three times a week. She went well with the Spitfire.

Our dates consisted of going to pubs and clubs. The *Rum Runner* in Broad Street was my favourite club. The music was great and the seating was arranged in huge old barrels.

The band at the time was the Kats. They played all the Beatles and any Liverpool sounds and also some Motown. Parking was a bit of a pain, so I soon approached the doormen and, after greasing their palms, became a bit of a VIP. The entrance to the club was down a long alleyway to the door. They let me reverse down and park outside the door.

You can't go wrong if you bung, the Old Man always told me. They watched over the car for me, and nobody else was allowed down the alley, so there was no problem getting out.

Rita was well impressed.

It's difficult to remember the exact timescale. After about twelve months we got engaged.

We started house-hunting and came across a newbuild development in Walmley, Sutton Coldfield. There were about twenty-five houses being built, by a small developer. We picked a plot that was only footings at the time. It was to be finished by September of that year, so we arranged the wedding at The Holy Family Church in Small Heath.

The house was four thousand, three hundred and fifty pounds. A lot of money at the time. But with my plans for the Merchant Navy gone down the old proverbial, I had the required ten percent deposit. The mortgage was twenty-seven pounds, four and nine pence a month.

The takings at the *Rum Runner* took a dive and the VIP parking was abandoned. Sacrifices had to be made. We watched all summer as the house rose from the ground and, in my usual style, I began bunging the right people. Always go to the top, the Old Man used to say.

I rooted out the site foreman, and soon there were a few added extras appearing in the house. Extra wiring, more plugs, wall lights, landscaping and so on.

The wedding went well. Plenty of beer and the usual wild west punch-up at the *Wagon and Horses,* Sheldon.

I sorted out the damage and no charges were pressed. Happy Days.

In the spring of the following year my daughter, Angela, was born. The eighth wonder of the world. I curbed my pub time to a good piss-up on Friday nights and Sunday dinner with my best mate and best man, Kevin. I don't think we ever missed a Friday or a Sunday in our lives.

In fact, I know Angela was born on a Friday because I celebrated the great event in *The Custard House.* Cheers.

In those days, they kept mother and baby in hospital for a few days until they were certain all was well. Fortunately, everything was and, when they were both home safely, it was the usual rounds of visiting and receiving visitors until everyone had seen the baby. Everyone was happy. Mom and Dad doted on Angela. Rita's family too.

And then, of course, there were my two young brothers, Ed and Phil, whom I have yet to mention. Only, I hasten to add, because they were too young to have figured in my life except as babies and young kids playing.

Phil was about fourteen by now and Ed was about seven. We were very close in our own way, but the big age gaps were difficult to bridge.

They would play a huge part in my life as they grew older.

Unfortunately, things ended tragically for both of them. And it was those two events that were to change me and my life forever.

With the wisdom of hindsight, I think the truth is that I got married too young.

Six

Fatherhood was great. Family life was great. I was enjoying it. The shop was doing OK. Getting a home together is great fun and takes time. You want nice things, so you work hard for them. Ang was growing fast and Phil was old enough to help in the shop. Ed was still a young kid. They both still lived at home with Mom and Dad.

Phil worked Saturdays and after school, and I was giving him a crash course in butchering. It was in his blood and he was learning fast. After work, we would go for a few pints and share a good laugh about the day and share any plans we had.

Sometimes, but not very often, we would play a round of golf. By now I was driving a brand-new MGB GT, that doubled up as a van. With the back seat down, it was amazing what you could get in it. I had a sliding, full-length sunroof fitted and he loved riding with me in it. We were growing into best mates, not just brothers.

Not being old enough to drink legally, in some pubs we had to sit outside in the car. Most we didn't. He liked a smoke - I've still got his last, unfinished pack of twenty Bensons after nearly fifty years.

His life was taking shape, and he was preparing to follow in my footsteps with his own butcher's shop. The Old Man bought him a shop in Edward Road, Balsall Heath. Like me, he was sixteen and shitting himself.

Knowing first-hand how it felt, I did everything I could to help him. However, he never had to worry about being too busy. It was a disaster. It was in the wrong place. Simple as that. We cut our losses and moved it on to someone who just wanted the premises as a base, and wasn't bothered about the passing trade.

Just as well.

We took a loss and almost immediately found a new place in Woodbridge Road, Moseley. Phil's confidence was severely dented and he was down. I told him truthfully that it was nothing to do with him. It was just a dead duck.

However, understandably, he was a bigger wreck this time than he was the last time. To make it worse it was coming up to Christmas.

Bloody turkeys!

We were all having nightmares. Three shops. Three nightmares. Normally, it would be every man for himself. To be fair, the Old Man had the busiest shop and the biggest problem. But, of course, he did what he could. Fortunately, Phil's shop was far busier than the previous dead duck and he had a good few orders.

In those days you had to dress all your own turkeys. They were dead, of course, but that was it. Heads on, feet on, guts in. They all had to be done by hand.

Picture this: 4.30am - up and off to the market, fighting off other butchers to get the birds you wanted. Then back to the shop. Work all day till closing. Then shut the shop and work in the back till about nine or ten prepping the bastards.

Me and the Old Man would work in our shops, then race over to Phil and help him. Then it was a few pints, a kip and start over.

It came and went. The door closed early on Christmas Eve, and then down the pub for a good few beers. It was over and, with a bit of luck, depending on what day of the week it fell on, you had two or three days off.

That Christmas, Phil bought Ang a battery-powered little dog, on a wire. She loved it and played with it all Christmas. I think she still has it now.

The New Year was never a big deal in our house. Mom and Dad had moved again, to Hatchford Avenue in Solihull. It was New Year 1973 and me, Rita and Ang visited Mom and Dad's to have a few drinks. I didn't know then that it would be the last New Year we would all be together.

• • • • • • • • • •

It was Monday. Half-day closing. The three of us, the Old Man, me and Phil, always met about 1.15 at a small pub, where we could always get the afters. In those days the pubs closed at 2.30. Phil had been taking driving lessons and had one booked for about 2.00 that afternoon.

He was obviously only drinking lemonade and, after checking his watch, he said his goodbyes and we wished him luck. Me and the Old Man had our quota and went our separate ways. He to Solihull, me to Walmley, where our usual habit was to have something to eat and catch up on our well-earned sleep.

Sometime around 4.00 in the afternoon the world went mad. The Old Man was on the phone. Rita was screaming and crying. I woke up in complete confusion. I ran downstairs. Rita thrust the phone at me.

"Phil's dead."

I snatched the phone. I couldn't understand what she was talking about. Mom was howling like a wounded animal in the background. I honestly can't tell you what was actually being said. It was all just like a Hammer House of Horrors film. All I can tell you is that I was in the car, with Rita driving, and we were on our way to the Old Man's, with Ang.

It's all a blur.

The police had been and gone. Mom really needed sedating. Poor old Ed, who was only eleven, was terrified. I think he was huddling with Ang, who wasn't yet five. I tried to get sense out of the Old Man.

Phil had been shot.

Shot on his driving lesson.

How can anyone get shot on a driving lesson? I was drinking with him at half one, for Christ's sake. There was a number I was to ring. I placed a call. The phone was answered immediately. It was Chief Superintendent Harry Robinson, the detective in charge of the case. Offering me his condolences, he tried to tell me clearly what had happened.

The driving instructor was in custody pending investigations and could I get to the city morgue to make a formal identification of the body, as they wanted to perform an immediate autopsy?

They needed a formal ID.

I truthfully told them that I was in no fit state to drive and didn't have a clue, anyway, where the morgue was.

He sounded a genuinely nice man. He told me not to worry about a thing. He was sending a car to transport me. Have a drink and sit and wait. We all sat around in the front room waiting. Me and the Old Man having a beer.

It was weird and unreal. I just can't honestly remember much about it.

It was dark outside, with only the glow of a street lamp in the room. But we didn't want the light on. I can't remember much being said.

I came out of my trance as a flashing blue light shadowed the room. There were heavy footfalls and a knock on the door. I answered it with the Old Man and, without a word, followed the policeman to his Panda car. He politely opened the passenger door and helped me in.

There were only the two of us in the car and silently, at great speed, we raced down the Coventry Road to Birmingham. We pulled in to the yard of an old brick building. I think it was by the Law Courts.

Waiting to greet me was a distinguished-looking man with a neat moustache, wearing a smart suit.

He put his arm across my shoulder and introduced himself.

"Hello Des," he said warmly, "I'm Detective Chief Superintendent Robinson. Call me Harry. I'm so sorry to have to meet you in such awful circumstances."

"Thank you," I said. "What is it you want me to do?" "It's not going to be easy for you, Des," he answered. I nodded.

"I'm going to take you into a room and ask you if you can identify the body I will show you. Have you seen a dead body before?"

"Yes."

"Take a few deep breaths and tell me when you're ready. I'll be holding on to you all the time."

"Thanks. I'm ready."

The door was straight off the yard. A big, heavy, old, wooden one. We entered a fluorescent-lit room in the centre of which was a gurney with a body on it covered completely with a sheet.

"Are you ready, Des?" I nodded.

A respectful-looking man in white overalls gently pulled back the sheet to just below the shoulders. Lying there before me in his cream leather coat was Phil.

I didn't collapse. I just burst into tears. They covered Phil back up and the only thing I can remember is leaning on the wall with my head in my arms with Harry supporting me.

They drove me back to Dad's.

Seven

Mom was inconsolable and totally hysterical. I told her and everyone else what Harry had told me.

Phil had started his driving lesson when his instructor, Keith Jones, told him that his trouser-belt had broken and did he mind stopping off at his house to get another one? Phil agreed and drove to his house in Armory Road, Small Heath. Jones had previously told Phil about a small gun collection that he owned and, according to him, Phil asked to see it.

They went in the house and, while Jones was showing Phil a homemade pistol he had, it went off, shooting Phil through the heart.

Apparently, Phil crashed back against the lounge wall and, as he slid down it to the floor, his last words were,

"You've shot me, you bastard." That's all I could tell them.

It was getting late and I had a three-year-old daughter to get to bed. Myself, Rita and Ang had to go.

I don't remember much else until the next morning, when it all started to sink in. I had a shop to open that my whole life depended on. Dad was in the same boat as me. We agreed to meet at the market.

He was having it harder than me. Mom couldn't understand how he could even think of opening up the shop. She just didn't comprehend the consequences of it all if the shop wasn't open. He tried to explain that meat wasn't like shoes and dresses. If you didn't sell it, you had to throw it in the bin. We hadn't got freezers like you have these days; if it wasn't fresh, nobody wanted it.

Phil's death was the only talk down the market. It was headline news in the *Birmingham Mail*. The newspapers constantly bombarded us for interviews, or at least a photo of Phil. We refused point-blank. The last thing we needed to see was Phil's face staring at us from a newspaper page.

Kevin was there for me. He popped into the shop a few times a day. He was a carpenter and worked locally.

After work we went for a few pints at *The Monica Pub*. It was local to the shop and his house, too. It was always busy with the after-work Irish community and we would get a warm welcome. Most were my customers.

Thursday evening in the pub I told Kev that I wanted to see this Keith Jones for myself. Needed to hear it from him. I had been warned not to go near him, as it would interfere with the case.

Without hesitation, Kevin volunteered to go with me. We decided on Friday night. It was common knowledge where he lived, as his driving school was locally-run.

I shut early Friday and went straight home to shower and change. I told Rita where I was going and phoned Dad. Nobody tried to stop me. We all knew that, sooner or later, it had to be done.

So sooner it was.

I called to Kev's house and went inside. His dear old mom threw her arms around me and his dad, brother and sister consoled me. Half an hour of sympathy and memory talk, and we left for Armory Road.

It was a grim area.

The BSA factory was at the end of a long, cul-de-sac road. Off it to the left were a few roads, all dead-ends, with terraced houses that had survived the war bombings.

The houses had small front gardens divided by knee-high wooden fences. We found Jones's, looked to each other, took a breath and knocked.

After a while. "Who is it?" After a while. "Des McGrath." After a while.

The door opened. After a while.

We went in.

There was no hall. We were in the lounge. Nobody had spoken. Looking around, I saw a thick smear of dark blood dragging down dirty wallpaper to the floor, where it had congealed.

"What happened?" I asked, breaking the silence like a whip-crack.

He told me, with his head bowed down, almost word for word what Harry Robinson had told me. He ended with Phil's last words.

"You've shot me, you bastard." Silence again.

We asked questions.

He answered them.

I stepped over to the bloody wall and stared at it, imagining the scene. I placed the palm of my hand on it and rubbed down the wall. I looked at Kevin, who flicked his head in the direction of the door.

We walked out into the black, gloomy night.

I didn't know what to expect when I went there, or what I was supposed to accomplish. But, when I left, I felt that at least I had accomplished something.

I couldn't for the life of me think of any other explanation why he would have shot Phil. And neither could the police.

After a thorough, six-week investigation, the final verdict at the inquest was misadventure. At last we could bury Phil.

Eight

We purchased two double plots at Robin Hood Cemetery in *Hall Green*. There was no question about who was to do the funeral. William H Painter. Michael Painter was Phil's best friend from childhood. He had lived next door to us in Sheldon when they were kids.

I was the only one who had seen Phil's body and, to be fair, once was enough for me.

However, Mom was insistent that she see him. Understandable. I called into Painters and spoke to the receptionist, who told me she would make the arrangements.

The next day I was in the shop when Michael visited me.

"Des, you've got to stop your mom from seeing Phil," he told me seriously. "Believe me, he looks nothing like Phil. She shouldn't see him like that."

He explained that he'd had four autopsies. Harry Robinson had even called in a London Home Office pathologist expert on firearms deaths.

"'The body looks nothing like Phil," he repeated emphatically. "You've got to stop her."

I was all in a panic again.

How do you stop your mother from seeing her son one last time? I finally convinced her but know, to this day, that she always blamed me for stopping her. I don't even know myself if I did the right thing. I'll just have to take Michael's word for it.

He's in no doubt.

The day of the funeral came.

I can only call it embarrassing. Mom would have no wake. The church was packed.

Loads of people followed to the graveside. How could you tell people that there was nowhere to go back to? Nothing. Not even at the house. What must people have thought? But the state that Mom was in, it wasn't worth the grief. We had to live with her.

The funeral car took us back to the house. Me, Rita, Mom, Dad, Eddy and baby Ang. We sat in the front room having a few drinks. I went back home later, with Rita and Angela, and sat taking stock of my life. I was twenty-four and, apart from two weeks in Jersey with the lads, all I could remember was work. Phil's death had shaken me. Until it happens to you, you don't realise the true meaning of "life is short". What had happened to the Merchant Navy? All I ever wanted to do, I remember, was see the world.

Well, I made up my mind. I was going to see it one way or another. It wouldn't be the Merchant Navy, so it had to be holidays. Holidays took money. Well, if that was what it would take, then I'd better set about making some.

Nine

I could never forget Phil, but life has to go on. I put a huge effort into the shop, increasing the takings dramatically. I didn't want to pay staff, so I worked flat out myself and managed with Uncle Bill and, later, his daughter Sharon, and anyone I could get. An after-school kid was always on hand to do all the cleaning and putting away, so there was a long procession of those.

Ideally, I wanted Ed, but the Old Man had already bagged him.

Across the road from the shop was an old, well-established, car repair business. The sign said 1930-something.

Arthur and Alice Drover were the owners. Lovely people. They had supported me with their custom from the day I opened.

Their two sons ran the place with a good few staff. Sid was the oldest, a couple of years older than me, and John was the youngest, a couple of years younger than me.

Sid was a bit of a playboy and had a lovely wife, Jennifer, whom I had always fancied with a major crush. She came into the shop a lot and we chatted loads. But for me, she could only remain long distance. She was devoted to Sid and remained so till he died around seventy.

After work, Sid, John and the lads always went to the *Custard House* pub for a drink.

They were always asking me to go with them so, one evening, I did. Soon I was going every day after work and having a few pints.

Sid and I were to become great friends. He and John both had caravans and would take them down to Aberystwyth in Wales. Abba, they used to call it. They were encouraging me to get one too and go down with them; I wasn't so sure. I'd never really fancied caravans.

However, the opportunity of a cheap van presented itself and I became the proud owner of what I can only call a vintage caravan. It had been owned by an old couple since new, who had decided they were getting too old for it. Although it was ancient, it was immaculately clean. Sid gave the mechanics of it the once-over and declared it fit and safe. I paid eighty quid for it. A price even then that was dirt cheap. It was only a ten-footer, so it was small enough that I could tow it with my MGB GT.

I bought a tow-bar for it and Uncle Bill fitted it for me. I was all set. I practised towing, with Sid's help, and soon felt confident enough to take my first trip. Secretly, I was looking forward to spending a little time in Jennifer's company.

Sid and John always went on Friday night. Because of the shop, I had to follow down on Saturday with Rita and Ang, who was four years old. There was just about enough room for her in the back of the car. It would never be allowed these days.

However, we found the site and parked up next to Sid and John, who was with his wife, Una, and a child about Angela's age. To call it a site was a bit of an exaggeration. It was a field with a brick-built toilet and washroom that seemed about a mile away.

We all sat around in the field on camping-chairs and tables. We cooked food and had plenty to drink and generally had a good laugh. The weather didn't look too good and the evening was getting cold. Sid and John had awnings on their caravans which, by the way, were brand-new. We all ended up in one of those.

As midnight came it was time for bed. It was starting to rain and we retreated back into our little tin can on wheels. The rain began to beat down relentlessly. It seemed to be turning into a storm. The caravan began rocking and the noise was deafening. The rain on the roof was like machine-gun fire. Angela was frightened and crying constantly.

I don't think we had a wink of sleep all night. Rita was complaining and really pissed-off. Not half as much as me! It felt like a boat in a bad storm. Anything loose was crashing about inside the small space.

By some miracle piece of engineering, there was a tiny toilet in the corner. Trying to use it was a feat in itself.

It was a long night.

Mercifully, the morning came and it stopped raining. I opened my little tin door and faced what can only be described as a sea of mud three or four inches deep.

Sid and John were already up and walking around in their wellingtons. I had none.

"Fancy a bacon sandwich?" Sid called over cheerfully.

Rita and Angela appeared behind and, together, we squelched over to our jolly neighbours. Jennifer and Una were both busy frying sausages and bacon and handing out endless sandwiches.

Things seemed better in the light of day and a belly full of grub and tea and coffee. We passed the morning chatting and, at pub time, Sid and John announced that they were going for a drink. Did I want to come?

I don't think Rita was too happy, but never mind. Me, Sid and John piled into Sid's Rover 2000 and he drove us to the local pub, *The Oak*. I'd heard a lot about the famous *Oak* and was surprised to find that it was a period establishment from a couple of centuries gone by, and still in its original condition.

A dump.

The locals all spoke Welsh and, though they seemed to love Sid and John, they made me feel as welcome as a fart in a confined space.

I was handed a broom and told that if the police came calling I was to start using it, as the pub wasn't supposed to be open on Sundays.

Great.

After our customary Sunday gallon, as we called it, eight pints, it was back to our luxury campsite. The girls were all cooking various dinners for their home-coming heroes who, as all heroes did, settled down in bed for a sleep, while the wives cleared up after the meals.

After a couple of hours' kip I said to Rita,

"I'm sorry about this. I don't want to spoil your weekend, but I think I'd like to go home. Do you mind?"

"Certainly not, darling," she replied. "Whatever you want."

I was going to say we packed up the van to leave, but the truth was I slammed the door shut and bade farewell to our disappointed friends and pissed off home.

I sold the van for 120 quid.

Happy days.

Ten

Post-traumatic stress disorder, I think they call it these days. I can't say it was as bad as that, but my caravanning nightmare in Welsh Wales had left me deeply scarred.

I booked a two-week package holiday to Spain. Never been before, but I was told the natives were a lot friendlier than the Welsh. However, that was over six months away.

There were a lot of old people in Small Heath; some had lived in their houses from the day they were built. One such person was a customer who lived with her daughter in a nice terrace up the hill from the shop.

She died suddenly and the daughter wanted to sell the house and move on. I had some spare cash and, cutting out the agents, she sold it to me quite cheaply. I offered to let her leave all the furniture if she didn't want it. She was delighted, as it saved her the hassle of getting rid of it.

I found myself the proud owner of a fully-carpeted and furnished house with everything from cups and saucers, bedding and towels, to a lawn-mower in the shed and all the garden tools. Not bad. With three bedrooms and a bathroom, I let it to four students almost immediately.

At about this time another butcher's shop came on the market. It was on the corner of Green Lane and Palace Road, not far from my shop. With a shop and living accommodation above, I snapped it up cheap, letting the shop and flat separately.

I was doing quite nicely and, while perusing the small ads in the *Birmingham Mail*, I spotted a registration number for sale, 52 DBM. I'd always fancied one, so I went after it and bought the old green log book for £25 and transferred my name onto the document.

I proceeded with the transfer only to find I had a big problem. You had to have the vehicle matching the log book, and the Department of Transport had to inspect the car before the transfer could go through.

The vehicle on the log book was a green Vauxhall Victor Estate. How the hell was I going to get around that? I could advertise to buy one and, even if I could find one, if it wasn't green I would have to have it painted.

All of a sudden, this cheap number was getting to be awfully expensive.

I used to drink quite often with Uncle Bill and his friend, Johnny McGoldrick - Mr Magoo. Mr Magoo had a small car-sales pitch in Erdington. He bought cars from the auctions and did them up and flipped them. Over a few beers we discussed my dilemma.

It was a tough one.

A few weeks passed and things were looking a bit hopeless. It was early one Saturday in the shop when the phone rang. Uncle Bill picked up and chatted, then called me over. It was Mr Magoo. I took the phone.

"Des?"

"Yes," I said.

"You'll never guess what's been abandoned on the waste ground at the back of the garage," he told me excitedly.

"What?" I asked.

"A green Vauxhall Victor Estate." "You're kiddin' me."

"Not a bit of it, Desi," he enthused in his broad Irish accent, "But you'll have to get it quick before the local kids get wind of it."

"Sure," I said, "I'll ring you back."

Myself and Bill exchanged some excited chatter, and I said, "I'm popping over the road to see Sid."

I was gone about fifteen minutes. When I got back we had a shop full of customers. Bill was busy. Getting behind the counter to help, I told him, "Sid's picking it up."

We were still busy serving when I excitedly said to Bill, "Have a look at that."

Sid arrived back at the garage in the breakdown truck. Smiling as always, a cigarette dangling from his mouth, he stepped from the truck and waved over the road, pointing to the back of the truck.

Dangling from the crane on its back two wheels was a bashed-up Vauxhall Victor Estate.

Calling into the garage, a couple of the lads came out and winched down the old wreck and pushed it up against a wall on the front of the office.

Now we were ready to rock. First thing Monday, Sid ordered some number plates. As they were brand new, Sid bashed them about and replaced the old ones.

The way it was in those days, the Department of Transport sent out a man to inspect the vehicle. He arrived one day with a clipboard and my log book, 52 DBM. I was right beside him as he carefully examined the car.

We've all seen it. Deep breath.

"You do realise, sir, that this is supposed to be a complete vehicle?" he began his summation.

"Yes," I said cheerfully, "Do you think two nice slices of sirloin steak would make it a complete vehicle?"

"I'm sure it would, sir," he said, signing off the paperwork as we crossed the road to the shop. I cut him two dirty great slices of the best steak and we both shook hands, happy. The transfer was completed.

SOH 995 H became 52 DBM. That, however, was only the beginning of my plan. I'd dreamed of owning a Jensen Interceptor. New, they were over five thousand pounds. To give you an idea of what that was in its day, an E-Type Jaguar was just over two thousand pounds new.

I'd been looking for one for a long time, when I saw what seemed a promising one. It was a 1969, one-owner car in fabulous condition. I can't remember exactly what I paid, but it was about two thousand for cash.

Shiny bright orange, with its 52 DBM plate it looked and felt brand-new. I loved it. And I soon realised it was a babe magnet. A young family moved into the road and started using the shop.

Geraldine was a beautiful woman, twenties, blonde, with long legs and a short skirt. She liked her meat. She used to hang about the shop when it was quiet. People started to notice and gossip. It didn't help Sid, John and the lads at the garage whistling and cat-calling across the road whenever she was in the shop.

After work, in the *Custard House*, it was even worse. "Well, have you, or haven't you?" the question would be. "No," I told them, truthfully.

"Well, you're barmy if you don't," said Sid. "Just give me half a chance."

Obviously, I'd never been unfaithful, but the temptation was too much. Geraldine was up for it, no doubt. She had a girl, the same age as Angela, so it was never going to be a love job. Neither of us wanted that.

But both of us wanted IT.

"Just as friends," we agreed, "Nothing in it."

Deal.

There seemed no point in delaying things. With both our bullshit stories rehearsed for our partners, plans were made.

We arranged somewhere safe where I could pick her up and take her to a pub where nobody would know us. We had a few drinks, but the excitement and anticipation of what was going to happen was too much for us.

I took her to a quiet spot that I knew from my courting days and parked up.

We only repeated the experience a few times; both agreed that it had been great, but it was getting too dangerous. We had one last fling and called it a day. She still continued coming in the shop and, apart from the odd sly wink, never spoke of it again.

But, like any drug, I had had a taste and wanted more.

Eleven

Holiday

It was summer 1973. Phil had been dead over six months. It was still very raw for all of us. The Old Man seemed to cope with it best. He'd had a tough life, and had dragged himself up from the poverty and hardship of his early days. As much as he mourned our Phil, he was hard and realistic. Life had to go on.

Of course, he was right. He passed his strength on to us. Eddy was thirteen now but had grown up fast. He was following in the family tradition. After school, and during the holidays, he worked in the shop and, like I'd taught him, he was stashing the cash. He had a savings account with *The Coventry Economic Building Society* and every week watched his balance grow.

There was a butcher's shop on the Coventry Road, Yardley, that a lot of people were watching. Butchers that is. It was trading at a fraction of its potential. It was old-fashioned and the old chap who owned it was way beyond retirement age. There was a *Dewhurst* butchers a few doors away that was thriving.

The Old Man was not one to hang about and lose it. As he would say, "You have to go out and get what you want."

To him, other butchers were the enemy. And you always had to steal a march on the enemy. And he did.

He approached the old owner and befriended him. It turned out that the shop was on a long lease, with large living accommodation over the top. Even better, there was an option to buy the freehold.

As it happened, the old guy who rented it had been considering selling for a long time but was daunted by the prospect of agents and solicitors and all the hassle that went with it.

Not a problem there. The Old Man had enough experience at buying and selling shops for the both of them. Offering to buy the lease and goodwill, based on the present takings, he offered a fair price cash.

The old chap selling didn't want anyone knowing that he was selling and, with the Old Man's help, the deal was quickly done.

Over a weekend, his name was over the shop. P McGrath and Sons.

Eddy had a couple of weeks off school sick and they opened it up together. It turned out to be a goldmine.

Within twelve months, *Dewhurst's* was shut.

However, Mom was a different matter. She couldn't get over her son being shot. There was nothing any of us could do to console her, as hard as we tried. She was never to get over it.

I'd found my way of coping.

 I shut it out.

Marrying so young, I believed I had to start living life more.

Maybe I was using it as an excuse to justify what the Old Man called my playboy lifestyle.

He disapproved of almost everything I did. Especially the Jensen Interceptor.

The time had come around for our first-ever family holiday abroad. In those days, flying to Majorca and staying in a nice hotel was still a novelty. I'd arranged for someone I could trust to run the shop.

The holiday started at the airport. The bar, the duty-free and the walk across the tarmac to what looked like an enormous plane. Angela was five. She had her own seat between myself and Rita.

The flight was an adventure on its own. Drinks from the trolley, duty-free fags and cigars and a bottle of gin and Bacardi.

Upon landing, the doors opened and the heat hit you like a desert wind. Fantastic.

Baggage collected, it was onto the coach and an exciting ride to Santa Ponsa. *Hotel Pisces,* up on a hill overlooking the unspoilt bay. The hotel was small, with a swimming pool and a nice bar. We were on full board. The dining room was spacious, with three long rows of tables, for four mostly. You were allocated the same table for the two weeks.

Obviously, to me anyway, the meals were all Spanish. It was all strange and new. It was all set menus. Some you liked, some you didn't. But we tried everything, and there was always plenty of bread and a sweet to fill up on.

The first few days were a bit strange, getting used to the place. The weather was great.

The beach was great. It was just the people I hated. Not the Spanish! They were great.

It was the British. Christ Almighty.

Moan! You never heard anything like it.

The food! You couldn't eat it. No roast beef, no sos and mash, or fish and chips. Pigswill.

And the beer. No mild. No bitter. No Guinness. Just Spanish lager that tasted like piss. And hardly anyone spoke English!

Hello!

It was my first Spanish holiday and even I knew it was a foreign country. They were seriously getting on my tits.

"How many people in England speak Spanish?" I asked some posh, illiterate prat who was a company director and had his own helicopter at home.

They were all company directors, by the way, with Jaguars, of course. They were all loaded and wanted to know what you did and had.

"On the dole, mate."

That usually got rid of them. "Peasants."

Well that sealed my agenda for the holiday. I found a little Spanish bar. The *Telebar,* it was called, where we made great friends with the staff and customers, communicating in sign language and bits of Spanish they were teaching us.

What fantastic people. The owner was an entrepreneur and was opening a brand-new terrace bar down by the beach. We were given invitations to attend the opening on the Saturday night.

We nervously showed our faces. Everyone was dressed in traditional Spanish costumes.

The music was Flamenco, with matching dancers. We were dragged in, when spotted, and treated like guests of honour. We were introduced to everyone. The drinks were all free. There was a long table of buffet food and everyone spoke perfect English. It seems they were all the well-to-do business owners in the area.

A fabulous night with the most gracious and hospitable people you could ever meet. It was that night that I fell in love with Spanish people and their culture.

They were to feature heavily in my life and writing.

Back at *Hotel Pisces* it was midnight and the bar was closed. Not a company director in sight. The waiters and bar staff were gathered for a drink.

The head barman was called Amador. His sidekick I'd nicknamed Charlie Chaplin. "Moustachio," Amador called.

That was me. They always called me that because of my moustache. "You want a drink?"

We'd made good friends with the staff, about the only people we ever spoke to. "Si amigo," I called back, "Bambino to bed first."

A few words to Rita and she took an exhausted Angela to the room. She'd be back soon. "Wadda you want?" Charlie Chaplin asked, as I joined the group. Amador and Charlie behind the bar. The rest of the staff around the bar stools.

"Cervesa La Grande, Gin tonica," I said, using most of my knowledge of Spanish.

"You want a Fokink Gin or a Gordons?" he howled with laughter, followed by everyone else.

I think they were all pissed, which was great, because I know I was.

It was a well-worn joke. There was a local gin, Fokink.

"Fokink gin, please." Local drinks were half the price and just as good.

With Ang safely tucked up in bed, Rita arrived and it was another start to the night.

It finished at 4.00 am.

Not bad really.

Six in the morning was not unknown.

Two weeks flew by. Nobody ever wants to go home, do they?

But with many new friends made - some would remain so for many years to come - it was back on the coach to our life as peasants, scorned by a coachload of company directors who would never be returning to the Godforsaken land of shit food and crap beer. Hanging our heads in shame for having a good time and liking the beer, food and people, it was back to reality.

Until next year.

Hey ho.

Twelve

The first thing we did on returning was book the same holiday for next year. Shit food and crap beer and all.

A few roads away, a house came up for sale. It was in a rundown condition, with two tenants. Upstairs was a large flat. The downstairs was like a rabbit warren and was inhabited by a totally bonkers old lady.

She'd lived there for years and was classed as a sitting tenant. Basically, she was there till she died. Nobody would buy the place. It was in a desperate state of decoration, but the building was sound.

The state was paying her rent and, at my age, I figured she would probably die before me.

I hoped so, anyway. But after what had happened to Phil, who knows? Nobody could get a mortgage on it and nobody in their right mind would pay cash for it with a sitting tenant. I made an offer for about a third of what it would be vacant.

I must have offered too much. They snapped my hand off.

Looking at it long-term, I'd paid cash and was getting two rents off it.

The old lady was really nice. Completely nuts, but harmless. Later on, I was to find myself doing errands for her. Hello! One time, when she was poorly, I found myself calling in the doctor and fetching and administering the meds.

With two houses and two shops to organise, I was kept quite busy. I was getting a little bored with the *Custard House* after work. Because of the shop you constantly get to meet new people. One of my customers, Alan, and his mother were chatting in the shop one day.

Alan was a car salesman at a large Ford dealership called Hangers, on the Kingsbury Road. He told me about a great pub, *The Norton*, that he and the other salesmen always used. I arranged to meet him there after work.

It was a great pub. Full of atmosphere, a lot of professionals and young people. Alan's friends were great and welcomed me into their company.

They were funny and lively and liked a drink. They only had two subjects, cars and women.

And I liked both.

They loved the Jensen and, to be fair, I think it played a big part in my instant acceptance. It seemed to open doors. It didn't go unnoticed by the girls, either. There were two in particular, who were barely eighteen and lived in Walmley, not far from me.

They were great fun to be with, fabulous to look at and great for my image. They were great friends and good company though, a few years on, Sue was to become more than just a good friend.

It was the New Year 1974. One year since Phil died.

Since our holiday in Spain, Rita and I had got into the habit of going out every Saturday to the *Buccaneer Bar* at the *Excelsior Hotel* by the old airport.

We used to meet the Old Man and a few friends, and always ended up having afters in the Residents' bar. We were inevitably talking about Mom. He explained how difficult it was at home with her. She was sinking into a depression and drinking too much.

He didn't know what to do; she made him feel guilty for being alive. Well, I knew that feeling. How could I go around buying shops and houses and go out drinking, while Phil was lying in his grave?

It wasn't fair on Eddy, either, living constantly in that atmosphere. But what was the solution? He knew his anniversary on the 15th was going to be a nightmare. So did I.

He'd thought of a possible solution and wanted to run it by me. There was a nice house for sale about a mile away. His idea was to buy it for her so that she could live in it on her own and he and Ed could continue living at Hatchford Avenue. They wouldn't be splitting up. They still loved each other. They just couldn't live together.

What did I think?

"If it works for you, fine."

And that's what happened.

• • • • • • • • •

Somebody had put a nail all around the Jensen. It was about March 1974. The whole of the car was deeply cut. It had to be completely resprayed.

I was gutted.

It was a real downer. It was Saturday, but Rita and I were not going out. Sid had lent me the works van to drive around in while he did the work. It was a battered old Ford Escort with an orange flashing light on the roof.

I drove it after work to *The Norton* to meet Alan. It was really lively and the jukebox was doing a good trade. After a few beers I was chilling out; Alan was talking to a fabulously beautiful girl who was eighteen and had just been crowned Miss Beautiful Eyes, Sutton Coldfield.

I don't know what the competition was like, but she definitely would have had my vote. As the evening passed, I seemed to be taking up more of her time than Alan. Apparently, she wasn't too interested in Alan's sales figures at Hangers. We were definitely hitting it off, much to Alan's obvious irritation.

About ten o'clock, Alan went to the toilet. I said to the beauty queen, "You don't want to be with him, do you?"

"Not really," she told me.

"Fancy going somewhere else?" "You bet!"

"Come on then."

And with that, we scarpered.

Weldall's van was parked close; we climbed in and I put on the spinning roof light. Just as Alan came out the door to see where we had gone. We'd both had a fair bit to drink and were laughing hysterically.

Alan stood, hands on hips, as I did a circuit of the car park, with lights flashing and my window down. I shouted something like, "See ya," and we were off up the road.

"Where we going?" she asked.

"A restaurant," I said. "You know a nice one?"

"There's a nice Indian in Erdington that I go to with my mom and dad," she replied excitedly.

"OK," I laughed, "Show me the way."

"Can I call in home first and tell my mom and dad where I'm going?" she asked.

"Yeah, no problem," I said. "Where do you live?"

"Walmley Road, it's on the way."

I knew where it was. She lived in one of the older, larger houses in the service road. Just down the road from me! Christ. I was so glad I wasn't in the Jensen.

We stopped outside and she jumped out.

"Come on in for a minute," she laughed, full of beans. "Say hello."

That's when I knew that I had to be pissed, because I went in and met her parents. I think she just wanted to reassure them what a nice person I was and how safe she was going to be.

I don't think it worked too well. Anyway, we were soon out of there and she directed me to a restaurant off Erdington High Street. It had an upstairs, and that's where we went.

We had a few drinks at a small bar as we waited for a table. She seemed to know everybody and had a contagious laugh. Her eyes were crystal-clear deep pools of soft blue, and huge. She was so funny and vivacious, we just never stopped laughing.

She told me honestly that she was having the best time in her life. The time fled and it was time to leave. We'd had a super bottle of wine, which she drank the most of. With me supporting her down the stairs I said,

"Shame to end the night now, just when we're having such a good time." She agreed.

At the back of the shop in Somerville Road was a small, comfortable bedsit. Mercifully it was vacant at the time and I took her back there.

The plan was to have some fun for an hour and get us both back home.

Well, you couldn't imagine our horror when we woke up in bed at eight o'clock on Sunday morning. And being Sunday morning in Small Heath, the whole population was streaming past the shop on the way to Mass at the Holy Family church. I explained that they were all my customers and I couldn't risk us being seen coming out of the shop.

"We'll have to wait until Mass starts and they're all in the church," I explained. "What are we going to do?" she panicked.

"We might as well get back in bed until they've all gone." "Might as well then, I suppose."

When the way was clear, we escaped to the van and drove off quickly.

"What am I going to tell my mom and dad?" she said, with genuine worry.

"If you have any brainwaves, can you share them with me?" I told her, equally worried.

"What am I going to tell my wife?"

There was a quiet country lane, Fox Hollies Road, at the bottom of our road. After I dropped Beautiful Eyes home I went and parked up. I needed time to think up some bullshit.

The only thing I could think of was that I stayed at Alan's. I didn't hold out much hope, but it was all I had.

Our kitchen was on the front of the house and, when I parked on the drive, Rita was in the window.

Shit.

When I got in the house the interrogation began. I told my story.

"You're a lying bastard. I phoned Alan last night, I was so worried about you," she began. "He last saw you leaving *The Norton*."

A full-blown row began and, to be fair, I deserved everything I was going to get. I'd messed up big time.

"OK, OK," I cried, putting my hand up. I might as well just tell you the truth and get it over with."

"Well?" she waited.

"I was pissed," I began. "What with the Jensen and that, I got well and truly smashed. I was driving down Fox Hollies Road and I thought I was going to wet myself. I stopped the van for a lag, got back in and fell asleep. When I woke up I panicked. I knew you'd never believe a story like that, so I said I stayed at Alan's."

"You silly prat," she said. "Why didn't you just tell me that in the first place? I know what you're like. I'd have believed that."

"Well I'm sorry, Reet," I apologised. "I just never thought you'd believe me." "Anyway, you shouldn't be driving with that much to drink," she scolded me. "You'd better be more careful in the future."

"You bet I will." Phew!

Thirteen

I avoided *The Norton* for a while, grabbing a couple of pints anywhere handy after work. I would get home early, have my dinner and watch some TV.

One evening about 8.30, the phone rang. Rita answered it. It was Sid Drover. I took over the call. Did I fancy a game of darts at *The Plough and Harrow* on the Chester Road? It was his local. I asked Rita if she minded. She was OK, so I washed and changed and went out.

Sid and brother John were there with a neighbour of theirs, Frank Deli, who was Hungarian. They called him "The Goulash Kid." He was a nice bloke, and our darts meetings became regular, Tuesday and Thursday.

The time was coming for our triumphant return to Santa Ponsa. Amador Garcia and Charlie Chaplin greeted us like long-lost friends. Angela was six, and far more able to enjoy things. We spent days by the pool, swimming and sunbathing for that all-important tan.

Most nights were spent in the hotel bar. It was there that we got talking to a really nice couple, Sid and Jane Pickering. Sid was a bookmaker. His life seemed to run parallel to mine. His Dad had about half a dozen shops in Birmingham and Sid, like me, had been brought up working in them.

Coincidentally, they were roughly in the same areas as us.

We got on really well and exchanged our details at the end of the holiday. Normally, you do that and never see the person again. Not so Sid and Jane. Soon after settling down again back home, we were meeting Sid and Jane for drinks and meals.

Sid was a larger-than-life character with the personality to match. He was short and fat, with tight, curly hair and glasses. He was an all-rounder. He could sing and tell jokes as well as any stand-up comic.

He introduced us to all his friends and we were building up quite a social life around them.

Some were couples, some were single. All were down-to-earth, normal people like us.

Sid was a born organiser. He was always planning something. A trip to *Hall Green* dogs was always good for a laugh. He organised cabaret nights at the *Night Out club* in Birmingham. There always seemed to be sixteen of us. We saw all the great acts of the day, Rockin' Berries, Hollies, the list was endless.

Sid was an expert on Chinese food and, as I'd hardly ever eaten it, he educated me. The *Shirley Temple* was his favourite restaurant. I don't think he ever actually had to pay. The business he brought them more than covered his food.

New Year 1975 was upon us again.

The dreaded New Year. Mom was living on her own in the new house, and Ed and the Old Man were in the old house and doing spectacularly well in the shop. Eddy was fourteen now and growing into a man. He had Dad's black, curly hair and was managing to get to the pub occasionally. Soon I thought we'd be going out together for a pint.

As usually happens in the New Year, talk got around to holidays. I'd intended to go back to Santa Ponsa but hadn't booked yet. Sid suggested making up a party. Did I fancy it?

Yes, I did.

There were twelve of us altogether, plus Angela.

We couldn't get in to the *Pisces Hotel* but, not far away at the bottom of the hill, was *Hotel Casablanca*. It was larger and could accommodate all of us. So, we booked it for the last two weeks in June.

I was still meeting the lads on darts night. However, I think we were getting a little tired of the darts. Frank, The Goulash Kid, had found a new pub at Stonnal, a small village just off the beaten track on the way to Brown Hills. It was only a couple of miles down the road.

We decided to check it out. The Drovers, Sid and John, went with Frank in his car and I followed in the Jensen.

The pub was called *The Swan*. An old building that was refurbished to a very high standard, with a large car park. It was busy with young people and had a great atmosphere. We liked it.

First time in a new pub, it's best to keep a low profile. We stood well away from the bar, talking together and having a good time. The bar staff were obviously hand-picked and well-trained, pretty, young girls. It was my round, and Sid came with me to help carry the drinks back. The locals seemed friendly and made room for us to get through. I bought the barmaid a drink and we carried our pints over to our spot.

We all agreed that we liked the place and decided to make it our new local and meet there in future instead of *The Plough and Harrow*.

Sid, John and Frank started drinking there most nights. They were getting well-known to the locals and, being friendly and easy-going, they were getting quite popular.

Sid had bought himself a nice Jaguar Daimler Sovereign, which went down well. I'd added Wednesday to my visits, making three nights a week.

We were on first-name terms with the staff and most of the regulars. As always, the Jensen guaranteed you acceptance and would always draw a question from someone, especially car enthusiasts.

I think it was April. The evenings were warmer and brighter. A lot of people sat outside, making a bit more room inside. Two classy-looking girls, about mid-twenties, came in and ordered themselves a couple of drinks.

"Haven't seen them before," I said to Sid.

"Came in a couple of times last week and over the weekend," Sid told me. "One of them is a local, married to a fireman. Lives in the houses just down the street. Her friend, the really nice one, visits and they've been coming in for a couple of drinks."

I was interested.

There seemed something a bit special about the friend. I tried not to get noticed, but I couldn't help looking at her.

I think she caught me and adjusted her stance at the bar so that her back was to me. Last orders came and went. The pub started to empty. The two girls drank up and left, crossing the road to a row of houses. I said goodbye to the lads and found myself thinking about the new girl all the way home. I never really believed in love at first sight, but there was something about this girl that had my heart racing.

I slept uneasily that night, with thoughts of her drifting through my mind. If there is such a thing as love at first sight, then I think this must be it.

Fourteen

The next couple of weeks were just the same as always. Sort the shops and collect the rents from the houses and check on the old lady tenant in Somerville Road.

She was all right.

Life was normal at home and everyone was happy!

I looked forward to my nights at the *Swan*. The fireman's wife and her friend were in the pub three or four times a week and mostly stood at the end of the bar, chatting and refusing any offers of drinks from the chancers and sniffers.

Apparently, they were getting a reputation as the untouchables.

They probably just wanted to be left alone in peace. After all, one of them was married and lived in the village. It was my turn for the drinks. Up at the bar I was standing next to the friend. I caught her eye, said hello and bought the barmaid a drink. I nodded to Sid and he came over and helped me with the beer.

I think, at least I hoped, that she was expecting me to say more. If she was disappointed or relieved I didn't know but, at the end of the evening, we all went our separate ways.

The next couple of weeks were mostly the same. We got on to first-name terms. She was Teresa, her friend was Ann. Sid and I used to break from John and Frank for a while and drift over to talk to them. We would offer to buy them a drink, but they would rather stay on their own. They both drank large vodka and tonics and could sink a few.

Teresa drove a new blue MGB GT. I told her about the one that I had had. I have never mentioned it before, but I wrapped it around a lamp post, seriously damaging it. The back axle had been badly bent and, after Sid had mended it cheaply, off the insurance, he had advised me to get rid of it.

That was the only reason I had sold it at the time.

Obviously, she knew that I was driving a Jensen Interceptor, and was curious as to what I did. I told her truthfully that I had a couple of butchers' shops and a couple of houses which I rented out.

Teresa told me that she worked part-time selling new flats that were being built in Streetly, on the Chester Road near Sutton Park. She invited me to call in and have a coffee some time.

I said I would. And did.

She was living in Gypsy Lane, Erdington, with her parents and brothers. At that time, she omitted to tell me that her husband was the drummer in a very successful band, Black Sabbath. They were having difficulties because of his wild life of drink and drugs.

I found all this out later.

It was May by now and only a few weeks until my holiday. As the time approached, I suggested meeting up when I got back, for a drink and chat. I told her that I was going to be really busy organising everything with the shops and probably wouldn't be around very much.

And that turned out to be very much the case.

The gang, all twelve of us and Angela, met at Birmingham Airport and, after the usual drinks and high jinks on the flight, we ended up at *Hotel Casablanca*.

The general rule was that everybody did their own thing. Couples went their own way and Pete and Brian, the singles, went theirs. Nobody was tied to anyone else unless they wanted to be.

However, you could guarantee that, by the end of the night, everyone ended up in Johnny Valentino's Bar. He was an aging Spanish film star and his bar was decorated with all his memorabilia. It was a nice place.

He loved seeing us arrive, as Sid would get the music going and sing away or tell jokes all evening. People were attracted in by the atmosphere and it would be packed.

There was a yard of ale on the wall and, as I was good at it, I organised regular competitions. There is a knack to it. Most people ended up soaked when the bubble at the end burst. That was disqualification. I made a few pesetas at that.

By the way, there was a good baby-sitting service at the *Casablanca*, so we could put Angela safely to bed before we started to party.

The days were passed around the pool or on the beach. I used to spend some time with Rita and Angela then slip away to a bar with someone.

Pete and Brian had brought masks, snorkels and flippers and were snorkelling out to the rocks on the outer limits of the bay. Liking the idea myself, I bought some equipment and went off with them.

From an early age I'd always been fascinated by the sea. I never missed any programme that had anything to do with underwater. Hans and Lottie Hass were my first introduction. Then there was Sea Hunt, with Lloyd Bridges. In my teens it was James Bond in films like Dr No and Thunderball.

The water was crystal-clear, with abundant fish and seascapes. The colours were fabulous.

I didn't know then that, six months later, myself and Brian would be diving buddies and visiting Stoney Cove in Leicester every week with Comdean Services, our new dive club.

And that next year we would be back in Santa Ponsa, diving in the Med with Escuba Palma every day of our holiday.

It was my first taste of the underwater world that would play a huge part in my life for the next twenty-five years, only to end in my retirement after what was to be my third near-fatal accident.

Fifteen

The holiday was over and we were back home.

And I found that the first thing I wanted to do was see Teresa. She'd been on my mind most of my time away. I think, if we were honest, from the day we first met we both knew that we were going to end up in a full-blown love affair.

How deep, strong and long no one would have ever guessed. The very first chance I got, I met her at the *Swan*. Of course, there was no romantic floating into each other's arms.

We hadn't even been out together. I was soon to put that right. I would pick her up from home and take her to a nice little out-of-the-way restaurant pub, *The Wilkin*, in Brownhills.

I can't, with honesty, remember exactly what we talked about, other than the fact that it was a very pleasant evening. She was good company and easy to talk to. We laughed a lot as I told her about my holiday and what a character Siddy Pickering was.

She would later agree with me about Siddy, when she met him. She told me that night about her husband, Bill Ward, and how they had been childhood sweethearts. She had some great stories to tell about the early days of Sabbath.

She was very close in those days to Thelma, who, she told me, was Ozzy's first wife, and Janet and Vic Radford, who played with them before Toni Iomi took his place.

Janet, apparently, was fed up with Vic and the hardship of being in the band and never having any money. She told Vic it was time to get a proper job and leave the band.

Janet and Vic were to become good friends of mine and we often went out for drinks together. I once asked Vic if he ever regretted leaving the band. He told me philosophically, "Well, obviously, I regret not being in the band but, in all fairness, I don't think Sabbath would have been Sabbath without Toni Iomi."

Teresa went on to tell me that after Sabbath's first huge hit, her life was changed forever. She loved the lifestyle it offered. They had a big house in the country and one of her biggest thrills was getting up in the morning and going to the local shop in her jeans and jumper in the Rolls Royce.

Funny, I thought.

However, with the band continually touring in Europe and the US, they saw less and less of each other. When Bill was at home he spent most of this time lying in bed, drinking and doing drugs.

She wasn't into any of that herself and it just got too much for her, so she moved back home to Erdington.

I told her truthfully that I was married also, and not unhappily. I had a good life. My problem was that I had married far too young. I felt the need to do a few of the things I'd missed out on in my youth.

I explained how my brother, Phil, had been tragically killed when only seventeen and that had really confirmed to me that life was short and I wanted excitement. I had missed out on the Merchant Navy and now wanted to see the world another way.

I said how on holiday I had enjoyed my snorkelling so much that Brian and myself had decided to take up scuba diving. She thought that was a great idea and said I should go for it.

After our meal, we sat in the lounge having a few drinks. I was partial to a cigar and asked if she minded me smoking. She said no, she rather liked the smell. She was quite used to the smell of smoke, adding that it was usually something a bit stronger that Bill smoked.

The pub called time and I took her back to her mom's. We chatted a while outside the house. She told me the hours she was working at the show flats in Streetly, gave me a kiss on the cheek and invited me to call in any time for a coffee.

I drove home a happy man.

Sixteen

Business was booming and I was looking to expand a little. Nothing big or risky, just something small and safe. I wasn't a great risk-taker.

"Little fishes taste sweet," the Old Man always told me.

An opportunity came to me. There was a small terrace house in Oldknow Road, not far from the shop. It was empty and fully furnished; not top of the range, but ideal for letting.

Someone had died in it and the person who had inherited it wanted to get rid. Quick and cheap.

Just up my street. I took a look at it and made a really cheap, cheeky offer, expecting to be driven up a bit. However, the seller was happy and, in no time, the house was mine.

There was a large hospital in the area and I soon rented the house to four student nurses. I asked a fair and reasonable rent, on the understanding that the nurses would take good care of it. They did. They were nice girls. They kept it clean and, if there were any repair problems, I took care of them promptly.

Everyone was happy.

The rent was always cash and helped with my expanding lifestyle.

• • • • • • • • • •

Taking Teresa up on her offer, I called in at the flats on one of my half-day-closing days. She was her usually bright and bubbly self. She gave me a tour of the show home and, after a couple of coffees, I arranged to meet her at the *Swan* that night.

I phoned Sid at the garage and asked him to give me a call at home to see if I could go out for a drink. At the predetermined time, the phone rang. I made no attempt to answer it, leaving it to Rita.

She began laughing and chatting for a few minutes then said,

"It's Sid; he wants to know if you fancy a drink with John and Frank."

Looking at my watch, I said, "Yeah, tell him I'll see him down there," adding quickly, "If that's OK with you?"

"Course it is."

"See him about nine." She told him.

I got washed and changed and went to see the lads. When I got to the *Swan* they were in a corner with the girls, happily chatting away. I got myself a pint and joined them. It was all very friendly and jolly and, when it came to closing, I said quietly to Teresa,

"Can you meet me in the car park of *The Plough and Harrow* on the way home?"

She agreed, and I got there before her. I sat waiting, with an eight-track on the stereo. It wasn't long before her blue MG parked next to me. She got out and into the Jensen.

We talked a bit and had a bit of a kiss and cuddle, but nothing heavy. She didn't seem to want any more and I didn't push things. It didn't seem to matter. She was just such great company.

She told me that, apart from Bill, she'd never had relations with anyone else. She'd dated no one since going back home. She'd been propositioned by loads of people, mostly in the music business.

"One rock star is enough for me," she said, with just a hint of bitterness.

She went on to tell me that she was in the *Opposite Lock Club* in Broad Street when Roy Wood's band started hitting on her. She gathered quickly that she was some sort of a prize.

Everyone wanted to bed one of Sabbath's women.

She stopped going to the clubs, preferring to hang out with her old friends like Ann. And it wasn't much better in her local pubs. Everyone knew who she was and it was like half the boys wanted to date her just to say they'd been out with Bill Ward's wife. Then there was the other half, who felt intimidated "Because she was Bill Ward's wife," and wouldn't go near her.

"No-win situation," I sympathised with her.

"Well not really," she said. "The fact is, I didn't want to go out with anyone. I just wanted to be left alone. It's a big enough mess as it is."

"And then I came along!" I beamed. "To the rescue."

"Yes. Well. We'll see about that, shall we?" she said, mocking me, with a huge grin. "I can't help thinking frying-pans and fires, with you."

I was deeply wounded. And let it show. I started the car.

"I was only kidding," she said, almost in a panic.

"I know," I said, laughing and turning off the engine. "Give us a kiss."

I think that was when we knew that we were girlfriend and boyfriend.

After that night, we kind of sorted out a routine. I started to meet her after work at *The Norton*. It wasn't practical to go out every night, but we still met at the *Swan* randomly.

Teresa instantly became a hit with Alan and the other salesmen from Hangers. Sue and Vicky were regulars in our company, as well as a neighbour of mine who lived across the road. Chris and Lyn Sassons were good friends of mine.

Chris was arrested by a team of armed police one night at the house. I watched the whole thing from my front bedroom window. I saw him dragged shirtless from the house and thrown to the ground in the street.

All very dramatic, over-the-top, Sweeny-style, I thought. Chris was sentenced to about four years for what was described as organised crime.

"Organised crime," he exclaimed about two years later, when he was released. "You've never seen anything so disorganised in your whole life."

You can imagine how miserable life can be for a wife while her husband is in prison. She was crazy about Chris and just dragged herself through the long months, hoping for early release.

There's not much anyone can do to help. I waved if I saw her and occasionally stopped for a chat if on foot in the street.

"If you're stuck for anything, give us a shout," I'd say, just to be polite really. I mean, what can you do?

Well, just before New Year, he was released for good behaviour. I was having a party and, after a few drinks, thought about Chris and Lyn. I walked over to the house and knocked. Chris opened the door.

"I'm having a party," I told him. "If you're not doing anything, pop over and have a drink." "Thanks," he said.

"Just come straight in; the door's open," I told him, and went back home.

Not long after, they appeared in the house with cans and a bottle. By all accounts, they had a great time.

It was about a year on from that when I appeared with Teresa in *The Norton*. Chris and Lyn were dominating the bar with Alan and three of the salesmen.

Since coming out of prison, Chris had opened two used-car lots. Economy Motors. He was doing well and did a lot of trade in deals at the right money from Hangers. Greeting us like great, long-lost friends, he bought us drinks.

It was a right party atmosphere, with our large gang dominating the bar. Chris took over the conversation with one arm wrapped around Lyn.

"I want to tell you something about Des," he started, a bit seriously, I thought. "When I was banged up for a couple of years, he was the only one in the street who ever talked to my missus."

I didn't know that.

"And when I came out," he continued, "He came over to the house and invited us to his New Year's party. I'll never forget that. He asked me to his friggin' party when everyone else in the street was crossing over to avoid me."

Everyone seemed impressed by the story, although I felt a little awkward with it.

Another couple of drinks arrived in our hands courtesy of the sales manager, Trevor Griffiths. "Cheers."

"And I mean this, Des," Chris said, from the bottom of his heart. "If ever I can do anything for you, you only have to ask."

I'd been thinking for some time about getting a little van to run around in. The Jensen was costing a fortune in petrol. It seemed like a good time to mention it, so I said,

"Don't think I'm taking advantage," I said, "But if you ever come across a nice little cheap van, I could really use one."

That's when Alan piped up.

"Tell you what; I've just traded in a Ford Anglia van," he began enthusiastically. "It's three years old, privately owned, with low miles."

"How much?" was the first thing I asked.

"It's on the pitch at 120 quid," he answered. Turning to Trevor Griffiths he asked, "That red Anglia van I took in PX, it's up for £120; how much can Des have it for?"

"What does it owe us?" "A ton."

"Trade it to him for the ton."

I was proud owner of a clean-as-a-whistle Anglia van that was to give me years of service at work and transport all my diving gear every Sunday to Stoney Cove.

Seventeen

We'd been seeing each other four or five times a week for six weeks. I had a very busy life.

Monday to Thursday, I'd see Teresa after work at *The Norton*. Friday it was me, Kevin and the Professor, Brian, but I still managed to steal half an hour on the way home from the shop to see Teresa.

Saturday and Sunday nights, Rita and I always met the Old Man and some friends at the *Buccaneer*, where I had a really nice arrangement with the head barman. A West Indian guy called Eric. He was a great mate, older than me, about forty. He was from Jamaica and lived with his wife and kids in St Benedicts Road, by the shop.

He called me Master. I called him Cousin. We used to have a great laugh winding up strangers by the bar.

"Cousin," I'd call. He'd come running. "Yes, Master."

"Can you see to a round for me please, Cousin?" "Sure thing, Master. Coming up, Master."

There were usually as many as about eight of us in a round. While serving other customers, Cousin would magically make a drink appear for everyone and, like the West Indies Trade Winds, he would disappear.

No money exchanged.

Of course, everyone else paid. My little arrangement with Eric meant him coming into the shop and going off with a bag of meat. At the end of the evening, I usually left stuffing a handful of Castella cigars into my inside pocket.

Uncle Bill was usually one of the crowd and once sarcastically asked,

"Is it my turn to buy a round, or is it your turn to get one for nothing?" Uncle Bill was like that. It was his kind of humour.

"Yours," I told him.

It was August. My birthday month. I was anxious to get some serious time with Teresa. It was time to step things up a bit. I was desperate for us to have a night alone.

A whole night.

I can't for the life of me remember the story I came up with. It was so long ago. It must have been good because, the weekend after my birthday, I was on my way with Teresa to the Holiday Inn, Leicester.

Two birthdays in one month. Not bad.

We checked into the room, which I must say was really nice. It was a case of unloading the little bit of luggage that we had and getting down to the bar for a few drinks before dinner.

Priorities.

Before we left the room, Teresa went all shy on me.

"I've got you a little present," she said, and rummaged in a small travel case. "I hope you like it," and handed me a card and two gift-wrapped parcels.

I wasn't expecting it. I took them, thanking her and opened the larger packet first. It was a box of fifty King Edward cigars. My favourite.

I told her honestly,

"You shouldn't have done that - they cost a fortune."

She shrugged sheepishly. I opened the smaller box and inside was a Ronson silver lighter. A French *Varaflame* model. It had a leather fitted case. I was speechless. I opened and read the birthday card and, before it all got awkward, she said,

"You like?"

And as I was about to answer she laughed, "Well come on then. I'm gasping for a drink."

Holding hands, it was straight down in the lift to the bar. There were vacant bar stools and soon I had a pint of bitter and a large vodka and tonic before us. Another couple came swiftly and it was time for dinner.

I was just about to sign for the drinks when Teresa looked horrified.

"Oh my God. Jeff Lynne's over there, he's just come into the bar. I can't let him see me." "Why?"

"He's a friend of Bill's."

"But you're not with Bill, are you?" I reasoned.

"It's one thing living at Mom's. It's another thing being in a hotel with another man."

"But…," I started.

"The divorce. Grounds. Settlement," she half-gasped, "And I'm sure you don't want to be dragged into the middle of it."

"Copy that!"

It suddenly dawned on me. No, I don't.

We had another drink with our backs to the room. In the mirrors behind the bar, we watched as Jeff and three other men got up and went into the restaurant.

"Well that's blown it," Teresa smiled. "Room service?"

I looked as miserable as any man had a right to be. Then perked up. Slipping the barman a pound-note tip, I asked,

"Can you send me a bottle of your cheapest champagne to the room, please?" "Cheapest!" exclaimed Teresa.

Winking at the barman, I said, "All tastes the same to me." "Me too," laughed Teresa.

"Give us a pint and a vodka please, mate, to carry up."

"Sure thing," winked back the barman. "Enjoy your evening." "It'll be hard," I said seriously, "But I'll give it my best shot." "Bottoms up. Steady as she goes. Sir."

"You got it!"

The room service was fantastic. I can certainly recommend the Holiday Inn. We were into a new phase of our relationship. And there were to be many more.

Eighteen

There was a diving school located on the Moseley Road, Comdean Services. I made some enquiries and found out they met for baths training on Wednesday nights. I arranged to meet Brian Perrie, my holiday snorkelling buddy, at the baths to speak with Tony Dean, who ran the club.

I went with Teresa and introduced her to Brian. They got on great. Brian was very laid back. He had long, straight, black hair and beard. I'm six feet one and Brian was about an inch taller than me. Some people called him d'Artagnan, like the cavalier. Personally, I thought he would have made a great heavy-metal rock singer.

He immediately began to call Teresa Fred, much to her amusement. "Why do you call her Fred?" I asked.

"Because I can't slip up in front of Rita or anyone," he told me, as if it made perfect sense. "If I never call her anything but Fred, I can't drop you in it."

Good tip!

The course consisted of six hour-long sessions in the baths, performing various swimming tests and practising how to use the diving apparatus. It was good fun, but serious stuff. Your life could depend on it.

The time came for our first open-water dive at Stoney Cove, a large, water-filled quarry in Stoney Stanton, Leicester. We had to be there at seven thirty on Sunday. As we had no equipment, Tony Dean hired it all to us. Dive gear is extremely expensive, so most people hire it until they are sure they want to take it up. You would be amazed how many people start and give it up after a few dives.

It's a dangerous sport and you need to be very fit to do it. It's not for the faint-hearted. It can be so cold in our waters that hypothermia can be a big problem. Another big problem is visibility. Sometimes it can be as bad as nil. It's not like these holiday divers in Malta and Cyprus and such places think it is. A couple of lessons in crystal-clear, warm water with Christos the Great and they think they're Hans Hass.

Stoney Cove. Welcome to the real world.

Brian had arrived at my house at about six thirty. I was taking the van. Brian got straight in and, wasting no time, I picked Teresa up from the pub car-park, where she'd parked up. It took forty-five minutes to Stoney. We paid our divers' fees and parked overlooking the water, in the huge but rough car-park.

Tony Dean and his French wife, Charmain, arrived in their stretch Range Rover. It had been cut in half and a middle section welded in. They brought with them four other newbies. Loads of other club members arrived in their own cars and, all in all, there were probably about twenty of us.

Tony checked our tanks and regulators for our supply, made sure our masks were secure and tight, then gave us a briefing. We were all to jump in and meet up on the bottom in about twenty feet of water.

We did. And it was bloody freezing. The water soaked into the wet suits and, according to legend, your body heat warmed it. Well, if it did, it was taking a bloody long time doing it!

Satisfied that we were all safe and ready, Tony signalled for us to follow him, which we did, keeping the man in front of you in sight at all times.

The water was clear; vis was about thirty feet. We passed the cockpit of an old airliner that had been sunk for the benefit of the divers. It was exciting stuff for a first dive.

There was good fish life; roach, perch and a couple of big pike. The seascape was good; rocks and swaying plant life. After three quarters of an hour we arrived back where we started, cold, but excited and exhilarated.

Out of the water, we changed in the open air and greedily drank the hot coffee from our flasks as we all chattered on with rattling teeth about the dive. We eagerly listened to the stories the more experienced divers told about their more adventurous dives, and realised how much more there was to Stoney Cove than we had seen. But we would in time.

We learned that the practice of the day was to meet at the *Holywell* pub in Hinkley, about ten minutes down the road.

It didn't officially open until twelve o'clock, but the back door was open to the divers from eleven. Those of us that had finished our dives set off to meet there. It was a great, friendly, old-fashioned pub with fantastic traditional ale. Brian liked his beer as much as me, and Teresa could match us pint for pint with vodka and tonic.

We drank and played darts with the locals until about two in the afternoon. I drove Teresa back to her car and Brian back to the house. Dinner was cooked and, as Brian went home, I tucked in heartily and went to bed, to get up fresh and take Rita out to the *Buccaneer*.

This was to become our regular habit over the next few months as Christmas 1975 rapidly approached. My first Christmas with Teresa.

I can't remember what we did, or all that happened, only that when it was all over and we were faced with the prospect of dark, freezing nights cuddling in a car, I knew I had to find a solution.

I knew I was in this for the long haul. So, I bought us a flat. It was a nice, modern new-build, only a few years old. It was a one-bed, lounge, kitchen and bathroom, with private parking at the rear. It was perfect. It was just off Erdington High Street, behind the church. We moved in at the end of January 1976.

Well, Teresa moved in. I was a frequent visitor. We fitted it throughout with carpet and Teresa chose all the furniture; with very good taste, I might add.

It was great. Instead of having to go to pubs every night, I could have a few pints and call around with a few cans and watch TV in the warm and in comfort.

We never missed a Sunday at Stoney and, after passing our trials quickly, myself and Brian were free to do our own thing exploring the Cove and building experience.

The annual holiday to Santa Ponsa was booked. It was the same twelve who had gone last year. Brian and I had our certificates to dive and, as the time approached, we were making plans for our two weeks' diving with the club by the hotel.

After a time, we stopped mentioning it as I could see that it was making Teresa sad. But it had been agreed that nothing could be done about it, and I would be back before she knew it.

She put on a brave face and handled it well. As the time got closer, I had to spend a lot of time organising the shops and houses. Our meetings were being cut shorter. I told her how sad I was to be leaving her for a fortnight, but what could I do?

Secretly though, I was really looking forward to a couple of weeks' holiday in the sun, with none of the complications of running two lives in parallel. Not to mention the diving. The time had come and I said goodbye at the flat Saturday evening.

And then, as I always find it best, I closed Teresa out of my mind and went to the *Buccaneer* with Rita and the Saturday gang.

She would still be there when I got back.

Nineteen

There was great excitement at the airport as we all met up in the bar. Rita and Angela were busy nattering to the girls. I was catching up with the boys. The flight and transfers went smoothly and, after settling in well, we went to Johnny Valentino's, where we got a fantastic reception.

Sid entertained the crowd and we had our last drink in the bar at the hotel.

First thing Monday, Brian and myself took the short walk to "Dive Escuba Palma," the shop located at a small hotel overlooking Santa Ponsa Bay. We made ourselves known to Raymond and his son, Danny, who owned the dive school.

They were very helpful and, after scrutinising our BSAC qualifications, told us that before we could dive we would need to purchase a Spanish diving permit. Raymond said he could organise all that in the afternoon in Palma if we paid him the money for it.

We paid up and then he told us the dive charge. It was 425pts a day which, in English money, equated to three pounds fifty. The price of a pint forty years later as I write. How much that is in today's money is anyone's guess. He took us to see his boat and generally fill us in on everything that we would need for the next day.

It turned out that he and his wife owned the hotel, and he invited us back to the bar to buy us a drink. Immediately, we felt at home and had great vibes about the dives to come.

After a few beers we made our way back to the *Casablanca*, where we excitedly drove everyone mad talking nothing but diving.

"God help us," cried Siddy Pickering, "They haven't even been in the water yet and I'm fed up of it already!"

We got the message!

I decided to take Rita and Angela for lunch. There was a great little bar restaurant overlooking the beach. The owner was a young Spaniard, Jaimi. We knew him briefly from last year but hadn't used it a lot.

The food was cheap and really nice. Jaimi and his Swedish wife were so friendly and welcoming that we ended up having lunch there every day, after my dive. We were, in fact, to return there for several years after and, when Angela had her first holiday without us, he was to appoint himself her guardian.

She was about sixteen and was staying at the *Casablanca* with her best friend, Sarah Ward.

One day she was passing his restaurant when a group of Spanish lads started hassling them. Jaimi was out in a flash, chasing them off and warning them to leave the girls alone. They were his friends.

I've never found out if they really wanted saving or not!

But they weren't bothered again.

• • • • • • • • •

We had all our own equipment, that we had brought with us from England. All we needed to hire were the air tanks. Raymond had them fully filled and waiting for us on the boat at 9 o'clock sharp the next day. The boat was about thirty feet, with twin Perkins diesel engines and was steered from the rear with a rudder.

Danny was dive leader that day, with Raymond as captain. There were about a dozen of us altogether. French and Dutch, mostly. Two were really hot babes. French girls, mid-twenties I guessed. Siddy Pickering had appointed himself dive master and, as it was OK with Raymond, was to come with us almost every day.

Danny told us we were going out a few miles to a small reef that had claimed victim a small fishing-boat. Great stuff! We kitted up at the dive site and Raymond threw the *anchor* over the side. When it had bitten, he told us to follow the rope down to the bottom, about sixty feet, and to always keep the *anchor* line in sight.

We paired up in twos and fell backwards over the side into the water. Following the line to the seabed, Brian and I checked all our equipment and settled our masks firmly to our faces.

The underwater world can only be described as wonderfully colourful and fascinating. There was an abundance of small fish and the sight of the broken, wooden fishing-boat was, for our first open-water dive, amazing. Sixty feet down gave us sixty minutes of bottom time without the complications of decompression. I guessed that was why it was chosen for our first dive. It was safe, and Danny could assess each diver and identify anyone who may look uncomfortable and cause a future problem. Everyone appeared fine.

I had my brand-new Niconos camera and was soon busy wasting my 36-exposure film. Checking our air gauges, we realised that we were using far more than we should. It was just the excitement of our first dive and a bit of slight anxiety. As we approached the red after about forty minutes, we swam back to the *anchor* line and pulled ourselves up to the boat.

Great first dive.

We started changing and, steadily, all the others returned to the boat. Danny last. Even though he knew we were all there, he made us stand still while he did a head count.

Standard safety procedure that he always insisted on. A diver could easily be left behind.

Always safety first. Reassuring!

Back on shore we all piled into Raymond's bar, drinking and swapping stories. Every dive is always different and one person can see something that none of the others has. That's what makes it such an interesting sport.

I met Rita and Angela at Jaimi's for lunch. They were as excited as me and wanted to hear, as much as I wanted to tell, all about it. After a great steak and a couple of freezing-cold pints of San Miguel, it was back to the *Casablanca* pool to sunbathe and play in the water with Ang.

Teresa had, of course, crossed my mind, but this was family time. I was a lucky man. Rita was a good wife and had done nothing to deserve being double-crossed. I had a beautiful daughter whom I loved very much, and a nice home life to go with it.

It wasn't their fault that Phil had been killed and scrambled my brains. It was just a phase I was going through and, when I grew out of it, I hoped I would get away with it and come out the other end in one piece.

Only time would tell. I was sure of only one thing. Whatever happened, I would take care of them.

• • • • • • • • • •

The underwater excitement was not the only excitement happening. Mark and Jill. The youngest.

Mark had fallen head over heels in love with a German girl, Sylvia. I can't remember where they met, but they were inseparable. Jill, of course, was devastated.

What a mess!

Although the girls did their best to console her, it didn't change the fact that she was now alone on holiday. It was really complicating things for everyone. As much as people sympathised, the fact was that we were all on holiday and it was Mark and Jill's problem.

Trouble was, Mark didn't think it was his problem. He was head-over-heels in love with Sylvia. To be fair, most of the lads could see why. You couldn't describe her as anything but drop-dead gorgeous.

Any one of us would willingly have helped him out of the hole he was in. She was great company, too. She could speak English and it was no hardship on us when Mark used to arrive with her at a bar when the lads were having a drink.

The nights were the biggest problem. Jill. There was no way that Mark, Sylvia and Jill could all be in the same company. Opinions were divided, but Sid was with me. There was no way I was going out and leaving Jill alone. If Mark wanted to be with Sylvia, then great. But not with us in *Johnny Valentino's*. So that's how it was. Jill came with us.

• • • • • • • • • •

The diving was going great. Everyone was different. We all had our own little dramas.

Me more than most. One day I leaned over the side of the boat to wash out my mask and a wave took it out of my hand. Raymond was laughing at my horror.

"Two-star diver no need mask." Well I did!

He let me sweat until we got to the dive site, before opening a locker on the deck and tossing me a spare.

Another day we were at the end of a dive. While I was prising an oyster off a wall of rock, I lost my knife and watched it disappear into over a hundred feet of water.

Shit, I thought.

Everyone else thought it was hilarious. Another dive, I forgot the film for my camera.

But then there was the most serious of dramas. Not so funny.

We were on a deep dive. 140ft. A hand grabbed my arm. It was Danny. He reached for my contents gauge, pointing at it. I was almost empty and we had only just started the dive.

He signalled me to go up with him. He made me hold on to his weight-belt, facing him. He started taking me up, watching my air all the time. While I still had a small emergency supply, he signalled for me to share with him.

This was serious stuff. A lot of trust is involved. He took a breath and handed the mouthpiece to me. I was to take a breath and pass it back. Then he took a breath and passed it back to me.

At seventy feet he signalled to stop. We adjusted our buoyancy with our life-jackets and he checked his watch. We had to decompress. We hovered at the depth and I left it all to Danny.

When he said, we started to rise again, sharing.

At thirty-five feet he stopped us again, checked the time and depth, and assessed our air supply. Satisfied, he waited. Then, at the right time, he took us all the way up.

Raymond was surprised to see us back so soon. Danny explained that, by chance, he had noticed air was streaming from the back of my tank and went to investigate.

It could have been fatal. If I had run out of air, I would never have made it to the top. If I had inflated my life-jacket for an emergency ascent, I would probably have burst my lungs and got the bends.

I like to think I would have saved myself. But I doubt it.

When everyone was back in the boat and having a laugh about it on the way back, Raymond said,

"I think for Desi it is necessary a nurse."

Very funny!

• • • • • • • • • •

It was nearly the end of the holiday. We were having a special party at *Johnny Valentino's.*

We all decided to dress a little smart. Brian didn't have anything you could call smart. With reservations, I loaned him a pale blue corduroy suit that was a favourite of mine.

The night was a bit more special because Valentino was cooking a huge pot of stew, consisting of squid and oysters that Brian and myself had caught over the last couple of days. There were various vegetables and potatoes bubbling in the pot and Johnny had brought a huge bag of French sticks of bread to just break up and dunk.

It smelled fantastic.

The jukebox was playing a selection of records, including some that Johnny had recorded a few years earlier, and we were all getting pretty drunk. Siddy was doing Frank Sinatra and customers were even standing outside on the first-floor balcony.

D'Artagnan, i.e. Brian, looked great in my best suit. Jill seemed to be having a great time, which was good to see, and Jane, Sid's wife, was drinking and laughing the night away with Rita. Angela had come out with us, as she often did, and God knows what she made of it all. With bowls of stew and lumps of bread in hand, Sid, Brian and myself found space outside on the balcony for air.

Brian was slaughtered and waving his arms about when he disappeared backwards over the balcony. Looking over, Sid and I were horrified to see Brian spread-eagled on the concrete below. Dashing down the steps to the pavement, we were amazed to find Brian sitting up and dusting himself down.

Anxious and relieved, we asked him how he was. He got to his feet and said he was fine and was going back to the hotel. With that, he staggered off.

We rejoined the party.

As the end of the evening approached, everyone started drifting back to the hotel. Rita went first with Angela to put her to bed. I said I'd join her in the bar in a while.

Eventually there was just Sid, Jane, Jill and me. Jill was wrecked. I said to Sid, "We'd better get her home."

Sid and Jane agreed. But Jill was having none of it. She wanted another drink and was starting to get agitated.

"Sorry, Des," said Sid, "But I don't need this."

Jane shrugged in agreement. She wanted to get back for a last drink with the gang. "But we can't just leave her here on her own," I protested.

"You do what you gotta do," was Sid's answer. "Look at the state of her. It's her problem, not yours. Sorry."

Jill was sitting at a table piled high with empty glasses. I sat on a bar stool with a fresh pint. I wasn't in great shape myself. Johnny Valentino was collecting glasses and clearing up.

"You can leave her here," he offered. "I can leave a table and a chair on the balcony." "Thanks," I said, thinking there's no way that's ever going to happen.

"Come on, Jill," I said, in my best jolly voice. "Let's get you back home."

"I want Mark," she slurred. "Where is Mark?"

"He's back at the hotel," I lied. "Get up. I'll take you there."

She staggered to her feet and, with her arm over my shoulder and mine gripped firmly around her waist, we faced the first major obstacle. The steps from the balcony to the street. Clinging for dear life to the side wall of the steps, we stumbled and staggered down.

I was gasping for breath as we got to the pavement. We stopped for a rest. Christ, I thought, how will I ever do this? Where's a wheelbarrow whenever you need one?

I had to think this through. The Old Man always told me,

"A little bit of brain is worth a lot of brawn."

I could see the hotel, it was at the top of a steep hill about half a mile away. A long half mile! At the rear of the hotel was a series of terraces that led down to the sunbeds on the beach.

I was about a hundred yards from the beach. It would be half the distance to drag her if we took the beach. I reasoned that, if I could get her to the beach beds, I could dump her on one and go up the steps to the hotel bar and get help.

I decided on the beach. I also thought that if she got too heavy I could drop her on the sand without hurting her. Decision made, I began what to me was like a Sahara Desert trek, carrying a wounded colleague.

What had seemed like a great idea at the time was springing leaks after a couple of hundred yards. The sand was so deep it was hard going, and Jill was stumbling and falling all the time.

I dropped her with a sigh of relief. I took a few minutes' rest.

Jill was quite a big girl. Not fat. She was tall and strongly-built. Not the sort of slight little girl you could pick up and throw over your shoulder.

"Come on," I urged her, "On yer feet." She staggered up and I dragged her nearer to the water, where the sand was firmer and easier to walk on. It was no good. I would have to try and carry her. I heaved her over on to my shoulder and, trying to balance her weight like you do a hindquarter of beef, I carried her off, occasionally having to adjust her by her backside so that she was evenly distributed.

I could see the beach beds ahead, but I knew I was never going to make it. There was an outcrop of beach grass nearby. I got to it and lay Jill down on the sand just below the lip of it. I lay down beside her, panting for breath. She started moaning for Mark. I wasn't in the mood for it. If I could have got hold of Mark at that moment, I could cheerfully have strangled him.

Looking on the bright side of things, as I've always found it's best to do, her little skirt had ridden up to her waist and I suppose it was the chill of the night that made it more apparent that she wasn't wearing a bra. I hadn't realised the full extent of my exertions until I was resting, and decided I needed to sit a little longer.

Snapping my mind back to the job in hand, I got her to her feet and, brushing her carefully down to get rid of most of the sand, we stumbled the last few yards to the hotel beach-beds. The mattresses had been removed, so I laid her carefully on the hard wood. I made her look decent and walked up the steps to the sound of a party.

Entering the bar from the rear terrace, I ordered a pint and said loudly,

"Jill's spark out on a sunbed at the bottom of the steps on the beach. Do you think you could manage to go and get her?"

And, if you think it couldn't get any worse, Brian was holding up the bar in a tatty old, torn, blue-cum-brown and black suit that looked as if it had been dragged across a building site.

And it had, apparently!

• • • • • • • • • •

The last day. The last dive.

Danny was in charge that day. Raymond wasn't feeling so good. It was a lot for him to do on his own and on the way back he asked me to take the tiller.

I loved it. Sitting up top at the back of the boat with the sun beating down on my well-tanned back - exhilarating!

As I got off, Danny was tying the boat off to the dock and announced to everyone, "Capitan Des. Two-Star captain."

We went as usual to the bar at his place for a farewell drink. I was disappointed that Raymond wasn't there. I met Rita and Ang at Jaimi's for lunch and said goodbye to him and his Swedish wife for another year.

Goodbyes are never nice, and we were all sad as we kissed and hugged for the last time until next year.

I was in my hotel room after dinner when Siddy banged for me on the door. Opening it, I said,

"What is it?"

"Raymond is in Reception," Sid told me. 'Where is Desi?' he keeps asking. He's come to have a drink with us."

Over the moon, I ran down to the bar where he was waiting with Brian for us. More farewell drinks!

A really nice man. We had hit it off from the very beginning. I thanked him for two weeks of wonderful diving and told him that I'd be back again next year, as I knew I would.

With the holiday over, my thoughts returned again to Teresa. She would have missed me madly and I also wanted to get back for a quiet night in with her in our little flat in Erdington.

But fate has a habit of turning your life upside down. Especially mine.

In the next chapter, you will hear how I swap my Jensen, with disastrous consequences, for a Triumph Stag. How, after a great dive at Stoney Cove and an even greater session at *The Holywell,* I get stopped by the police and lose my driving licence for a year.

All of which would ultimately lead to a two-week adventure, diving in Barbados with my mistress, Teresa.

Happy days!

Twenty

It was July 1976. I had a brainstorm, as I often do, and made an impulsive, costly mistake.

I sold my Jensen and bought a beautiful, white Triumph Stag. I fancied a convertible and was taken by its fabulous new design.

Proudly, I took it to Stoney Cove. I had collected it on the Thursday and, as they were so new and few, it caused a great deal of interest with my fellow friends and divers. Everyone wanted a look and I was as proud as a peacock.

I had a great dive at Stoney and it was a lovely day. Brian had lost a bit of interest and I had found a new diving buddy, Terry. It was working out a lot better for me, because there was no picking-up and dropping-off to do with Brian.

Terry lived in Warwick and used to meet me at Stoney. That meant Teresa and I were able to be together alone.

Terry was never late, and we always got there at seven thirty so were one of the first in the water. That was really important, as the bottom at Stoney was very silty. Early on, it was clear as a bell and great for my photography. We were in and out before the visibility was destroyed by two hundred or more thrashing fins.

That day was just so good.

We found five huge pike that seemed to me to be posing for the camera. I got some great shots. Then we came upon a huge shoal of perch, then roach and, the cherry on the cake, a family of crayfish.

I used the whole of my 36-exposure film and was delighted. I would have stayed down all day if I had the air.

We hauled ourselves out of the water and, as we were changing, Teresa was cooking breakfast on our camping stove. It always drove the other divers mad. Bacon, eggs, sos and tomatoes with a loaf of fresh bread.

Excitedly, we told Teresa all about the dive, as she fended off the other divers all trying to scrounge a sandwich. No chance. I couldn't wait to get to the *Holywell* for a pint of Marston's Pedigree.

The locals always enjoyed our stories over a game of darts. Terry only stopped for an hour or so and got off home about one o'clock.

Me and Teresa stayed for afters and decided to stop at the *Wilkin* in Brownhills for food on the way home. They weren't open when we got there, so we parked the Stag and had an hour's kip.

Nicely refreshed, we were in the doors as they opened. After a nice meal and a few more pints, I set off for home. It must have been late, because it was getting dark.

My heart stopped as a blue light began flashing behind me. I pulled over and wound down the window. The police were friendly and polite.

"Do you know why I pulled you over, sir?" the officer asked. "No, sorry," I answered.

"Your nearside rear tail-light is out," he informed me.

"I'm really sorry," I said sincerely. "I only picked the car up on Thursday." "That's bad luck, sir," he said sympathetically. "Have you been drinking, sir?"

"I've had a couple of pints," I lied. "I've been diving at Stoney Cove and was on my way home. Was there anything wrong with my driving?"

"Not at all, sir," he replied. "I've been following you for about a mile and your driving was fine. However, I'm going to have to ask you to give me a breath sample."

He asked me to get out of the car and get in the police car. I blew positive, no surprise there, and he informed me that he would take me to Brownhills Police Station, where I would be asked to give another sample. His colleague would drive the Stag.

Teresa came with me in the patrol car to the station and, after blowing positive again, I was escorted to the cells. I have to say that they were very sympathetic and friendly. They left the cell door open and brought me a hot drink. Teresa was chatting away in reception.

It felt like I was playing a part in *Carry on Constable*. I found myself wandering out of the cell to Teresa in reception.

"Excuse me, sir," said the sergeant behind the desk, "You're supposed to be under arrest.

You shouldn't be wandering around the station." "Sorry," I said. "What happens next?"

"I can't let you go, sir, until you blow a clear result," he grinned, almost laughing, "Which by my calculations, looking at your reading, could be about midday tomorrow."

"Could you call me a taxi?" I asked. "Where do you live, sir?" he asked. "Sutton Coldfield," I told him.

"That's going to be very expensive," he said. "I'd have to be certain that you had the means to pay for it."

I stuck my hand in my pocket and pulled out a wad of about two hundred and fifty pounds. I showed him.

"Well, I can see you have the means to pay, sir," and, with that, he called me a taxi.

I can't imagine that things are the same today and I've no wish to find out. I told them I would collect the car tomorrow afternoon. They told me that I would have to pass another test before I could take the car and reminded me to get the light fixed.

I said I would. The taxi came and we said goodbye. It dropped Teresa at the flat and took me home.

Rita was waiting. She was not happy!

I had to go over the whole sordid tale again, in almost every detail, and collapsed exhausted into bed.

There was no market on a Monday so, a little later than usual, I opened the shop. It was always quiet on Mondays, thank God, as I spent most of the morning recounting the details to my mom, my dad and Sid Drover and the lads at the garage.

And the worst part of all was the tail-light story. It would have been easier to take if I'd had an accident. But a bloody tail-light.

I cursed the Stag.

I closed early and went home in the van. Rita had her own car, a Mini. She drove me to Brownhills, where I collected the Stag and she followed me back.

I parked the Stag on the drive.

And we went down the pub in the Mini.

About a month later I pleaded guilty and, after throwing myself upon the mercy of the court, I received what the magistrate said was a lenient sentence.

A one-year ban.

Normally, they said, in that court I would have received eighteen months. But, due to the hardship it was going to cause me and my charity work, they had decided to give me the minimum sentence.

Happy days.

Twenty-one

It's in times of trouble and strife that you think everyone will come to the rescue. Think again. The first few weeks I took it for granted that there would be no shortage of lifts to the pub and home again.

At first, I thought that it was just that nobody was going to the pub after work. Or they were doing something else. Then I bitterly found out that it was every man for himself. In fact, it seemed to me that people were gloating and wallowing in my downfall.

I was proud and would never ask for a lift. I knew it was not far out of certain people's way to give you a lift home, but I was never going to say, "Are you going my way?"

My main problem, however, was running the business. That's where the money came from for all the cars and holidays. The biggest problem was the market. Rita took me Tuesday and Thursday in the van for the big bulk of stuff. If I needed any small odds and ends, either the Old Man dropped it off or the market delivered it. Saturdays, Rita took me to the shop. The rest of the week I managed on the bus.

It worked. I'd get through this somehow. I'd show them. I was beginning to get a little bitter. I shouldn't have been. It was my problem. Nobody else's.

Well there was one person I could count on - Teresa. And boy, was she there for me. She picked me up every day from work for a drink at *The Norton*, or took me home and picked me up later for a drink and an hour at the flat.

A big worry was Sundays at Stoney Cove. I'd barely ever missed a dive. She had that covered, too.

She sold her MG and bought the most useless car ever made, an Austin Allegro. It was a gold colour, reg number LOF 1OP. And that's what she called the car, - Lofiop. It had plenty of room for all my diving gear and cooking stuff.

She used to pick me up at the top of the road at 6.45am and drop me back at the top of the road in the afternoon. Then she used to go home and drag all my gear up to the first-floor flat and wash and dry it all for me the following week.

The Stag was sitting on the drive with nowhere to go, so I sold it. Feeling a bit sorry for myself, I went to the travel agents with Teresa and booked a two-week holiday, to Barbados in October.

There was a dive club that operated from *The Coral Reef Club* hotel in St James, on the west coast. The hotel was too expensive to stay in, so I rented a one-bed bungalow not far away. Included in the whole package was a Mini Moke. All we had to worry about was food and drink.

It turned out to be a smart move, going self-catering. The price of food and drinks in the hotels was off the scale. There were shops not far away and a nice little restaurant on the bungalow complex, if you wanted a treat.

As usual, having booked the holiday on the spur of the moment, I now had to figure out how I was going to get away with it. I organised a butcher to do the shop. Rita and the Old Man would order and collect the meat from the market.

Rita obviously wanted to know more about my friend, Glen, whom I was going with, so I filled her in. She had heard me mention his name a few times, as he was a regular at Stoney Cove and the pub afterwards. What he did for a living was sketchy, but he always had plenty of money and the cost of the trip was no problem for him.

Rita was satisfied and things were slowly falling into place. We were flying from London on Saturday. We were taking the train to London, then the Underground to the airport.

The problem was that Rita was insisting on taking me, with Angela, to see me off on the train at New Street on the Friday evening. Glen and I were staying overnight at the airport to make the journey easier. Truth was, I was staying at the flat with Teresa Friday night and travelling to London on Saturday morning.

Rita waved me, with my suitcase and diving bag, goodbye as the train pulled out of the station. I had an extra ticket in my pocket that I had used to board the train. It was for the NEC, where Teresa was waiting with the car to take me to the flat.

The next morning Teresa's brother collected us and took us back to New Street, where I had started from the night before. By late afternoon we were on our plane to Barbados.

Happy days.

Twenty-two

There's nothing worse than being stuck on a plane for hours with a nuisance passenger. Teresa had the window, with me in the middle. Next to me in the aisle seat was a young girl, early twenties, beautiful and travelling alone.

Normally, that would have been enough of a distraction in itself. The fact that she was wearing a sheer black see-through blouse, with no bra, made it even worse. She was half-caste, with light-brown skin and very dark nipples.

Teresa was not happy!

But, to be fair, there wasn't much I could do about it, was there? And, after all, she hadn't got anyone else to talk to. By nature, I've never found it easy to be rude to someone, no matter how much they may have deserved it. I'd been brought up to be polite. So, I didn't know what to do when she dozed off with her head on my shoulder.

I'd never seen Teresa snarl like this before. "Comfortable are you there, Des?"

"It's not my fault."

"You don't have to keep staring," she seethed.

It was then I realised that I was, indeed, fascinated by her slim, barely-covered, bare legs.

As I looked away, I found myself distracted by the contours of her sheer blouse.

Christ. I don't need this, I thought, barely managing to suppress a grin. "Look," I said, "It's going to be a long flight."

"Yes, it is," she quietly snarled between gritted teeth. "You're telling me." "Fancy a drink?"

"Two large vodkas."

I pressed the call button.

Fortunately, we had a refuelling stop at the Azores, a group of Atlantic islands. They looked beautiful from the air as we were coming in to land. Understandably, the lovely girl in sheer lace top and pelmet-length skirt wanted to get a look at them also. She asked if I minded her stretching across me to try and get a look out of Teresa's window.

Of course, I reluctantly had to say yes and suffer a close-up of her cute little breasts hanging below my head. She seemed to be as excited as I was, enthusing about the view. She brought to mind the old music-hall joke.

"Is that a gun in your pocket or are you just glad to see me?!"

To be honest, I was glad to see her. You'd never believe me anyway, if I said anything else.

We landed and it was optional whether you got off the plane or stayed in your seat. Most people got off. It was quite warm as we climbed down the steps from the plane, and I had a good idea what the main topic of conversation was going to be as we reached the terminal building.

It was a very basic structure built of wood. There were a bar and rows of basic seating, occupied mostly by peasant-type people. I got two bottles of beer and a vodka and tonic for Teresa.

Sure enough, the conversation was as predicted. I just fielded it as best I could. "How would you like it if I was wearing something like that?"

"You wouldn't be. I wouldn't let you."

"She's practically thrown herself at you." "Well, you can't blame her for that, can you?" "She's sprawled all over you."

"There's not a lot of room." "No bra with a top like that!"

"I thought it was quite a nice top." "Well you would, wouldn't you?"

"It's all right for you. I'm the one that's got to suffer it for another four hours."

That was the rough gist of it. I just switched off. Silently loving it. It was time to board the plane again and suffer a few more hard hours.

We landed in Barbados in the dark. I remember the coach trip was a loud cacophony of crickets and bird sounds. The rep dropped us at the bungalow complex and Reception directed us to ours. There was a basic welcome pack of bread, fruit, tea, coffee, milk and sugar.

The bar, which was outside by the pool, was closed, so we had a snack and a duty-free nightcap. We went to bed and woke in the morning to the most beautiful island day.

The first thing we did was go to Reception to see about our Mini Moke. It was in the car park. After sorting out the paperwork, I found out where the shops were and set out to get some supplies.

Almost all the meat was frozen. If you wanted fresh, you had to go to the market in Bridgetown. I bought mostly lamb and pork chops that looked OK and a big bag of chicken portions. With bread, spuds, eggs and bacon, I figured we wouldn't starve.

We took it all back home and, after storing it all away, set off to find the *Coral Reef Club*. It wasn't hard. It was a couple of miles down the primitive coast road.

Barbados was not turning out to be the island paradise I had dreamed of. The local homes seemed to be row upon row of wooden sheds on either side of the road. They were of various lengths by extending from the rear. It appeared to me that, the more people there were in the family, the more sheds were added on to the back.

The bars I passed were bigger, longer-fronted sheds with a drop-down hatch that revealed a few bottles. There were a couple of rusty old tables and chairs outside but, wisely, I thought, the customers sat on the ground outside, with a bottle in one hand and a rather fat cigarette in the other.

I made a mental note to cross local bars off my to-do list.

The Coral Reef Club was beautiful. On the beach, with a tropical beach-bar atmosphere at the back, we sat down at a table on the sand and ordered drinks.

I asked about the diving and was told that today's group had gone out and would be back in about an hour. The hotel was not all that large but was certainly luxurious. After receiving the bill for the drinks, I was glad I'd gone self-catering.

I made a note to stop again at the shops and load up with beer.

After an hour or so, I spied a couple of flat-bottom boats with small outboards chugging slowly towards the beach. There were about six divers in each boat and, as they scraped up onto the sand, one jumped out into the surf and dragged them away from the water.

Hauling their equipment up to the bar area, the usual diver banter began as they found tables and ordered beers. The last man in was an old geezer who looked the spitting image of Mr Pastry. Les Watton, owner of the dive school for the last twenty-five years. Proud to say he never lost a diver.

That was reassuring. I thought he must have begun very late in life. He was about five and a half feet tall, skinny as a rake, with knobbly knees and a baggy pair of vintage khaki shorts.

I left it a while before introducing myself. We shook hands. He'd been expecting me. He gave me a rundown of the operation, which was all very casual. It was forty Barbados dollars a day, which worked out at thirteen quid. OK, I thought.

There was a shed further down the beach at the side of the hotel that served as the dive shop. All I had to do was turn up at nine in the morning with forty dollars and my equipment, and he would supply the air tanks. No probs.

After mixing in with the divers for a bit and another drink, we set off back, stopping for beer. Our balcony overlooked the pool and had the sun in the afternoon. We sat sunbathing and drinking, and I decided to eat in the poolside restaurant that night while we still had some money left.

So that's what we did. Showered and dressed, we ate battered flying fish and drank cocktails in the moonlight until closing time. Then we wandered hand-in-hand back to our bungalow to spend the rest of our first day drinking vodka and Bacardi on the moonlit balcony, dreaming contentedly of the life we could have together and the excitement to come over the next two weeks, diving in our island paradise.

Twenty-three

The talk of the island, I was to learn the next day, was the fate of the Cubana Airways plane and all its passengers that had crashed soon after take-off. The belief was that it had been the victim of a terrorist bomb. To the best of my knowledge it was never found, nor any of its passengers.

The water where it came down off the island was deep. Miles deep.

Bang on time, I met with Mister Pastry and seven other divers at the *Coral Reef Club*.

Collecting our tanks and heading for the two flat-bottom boats that were our transportation to the inner reef, we chattered amongst ourselves how good it would be to find a piece of wreckage.

Teresa settled down on the beach with a good book amid the swaying palm trees and watched as we chugged slowly out to sea. Mister Pastry was giving us a guided-tour chat about what to expect and paired us up into twos. "You can't dive alone, first rule." Then, with what I guessed was his stock joke, he told us,

"And finally, if you run out of air, it's time to come up." We all laughed dutifully.

I teamed up with another first-timer, Alex, an American. He seemed all right but a bit of a know-all. Pastry had told me I wouldn't need a wet suit - the water was 70°F - so I'd left it with Teresa and, for the first time in my life, I dived in my Daniel Craig Speedos, with my camera dangling from my neck.

Alex was armed with a bang stick, a three-foot pole that was loaded with a shotgun cartridge. Apparently, if a shark got too interested in you, you could stab it against him and it blew a hole in the poor sod.

I preferred the old-fashioned method. Flip to the side of it, roll onto its back grabbing it by the dorsal fin, and plunging my knife into the soft flesh of its underbelly.

Each to his own, I suppose.

Considering that it had only been about twelve months since the film

Jaws had been released, everyone was very shark-conscious. One silly prat was hammering on about how much he wanted to encounter a shark.

I was hoping to God that I never saw one at all. But at least I was prepared. However, just the same, I thought it a good idea to hang close to Alex, just in case! In case, that is, of course, that he missed with the bang stick and needed me to save him.

As you would expect, the dive was amazing.

The coral was spectacular and the reef fish abundant, every size, shape and colour imaginable. The vis was about a hundred feet. I'd never had vis like that in my life. Swimming to the bottom on the outside of the reef, the depth was about a hundred feet. Looking out beyond, there were thousands of miles of sloping, flat sand that hid only God knows what predators and wrecked ships of all sizes and types, by the thousand.

An awesome thought.

Just within our vision swam a school of lazy-looking, uninterested barracuda. Swimming slowly back up the reef, I saw the head and about two feet of moray eel poking out of a small hole. I got close and took a photo that, until this very day, hangs behind my bar at home.

If I never had another dive for the rest of my life, I think I can safely say it would have been enough.

Of course, it wasn't going to. There would be hundreds more to come.

Back at the *Coral Reef Club* we pulled the tables together and drank beer on the sand beneath the palms. Teresa laughed, joked and joined in the fun like one of the lads. She could drink vodka one on one with our pints. Her blossoming tan blended nicely with her yellow bikini.

I think she was more popular than me!

Over the two weeks, many different divers came and went. All, in the short time that we knew each other, were to become friends.

Harry, in particular, was a four-day friend. He was older, in his thirties, and was a pilot for an airline in Rhodesia. He'd already had a week in Antigua and was island-hopping. I asked him what it was like there.

"It's a slightly bigger shit-hole than this place," he told me.

I was relieved to hear that I was not the only one of that opinion. One day, after a dive, we were talking about flying. He had an observation about that, too. Harry seemed to have an observation about everything.

"The darker the skin, the more dangerous it gets," he said. One day, he didn't feel too well.

"Too many rum punches," was his excuse. He kept Teresa company while I dived.

When I got back, and after the usual banter with the lads, he insisted on buying us lunch in the hotel. God knows what it cost, but he was most insistent. We had kingfish and a lovely white wine that he recommended.

I'm not too sure what the drink-drive laws were in Barbados at the time, but I hadn't got a licence anyway, and I wasn't too bothered about being banned from driving in Barbados. We made it back to the bungalow and sat on the balcony in the sun, overlooking the pool and bar, and carried on drinking vodka and Bacardi till about midnight.

Good dive-training, eh?!

With Harry gone, I teamed up with another American, Dick. He was an obstetrician with his own practice. He was on holiday with his wife, who preferred to stay at their hotel while he dived. He had a terrific sense of humour and we got on great.

He also had a camera, so it was great picking his brains. He had far more experience than me. Like Harry, he was a lot older than me and had dived all over. Hawaii and the Cayman Islands were the best in the world, he told me.

I let him lead. A good idea. He could tell by the terrain where to find the best fish. It wasn't long before he urgently signalled me to stop and be still. Gingerly he moved forward, me following. He stopped and pointed.

In a large hole in the reef was the biggest fish I've ever seen. It was huge, with a big mouth and black-and-white spots. A grouper. We watched it for a few minutes, adjusting our lenses on the cameras. As we moved he looked at us and, with one effortless move, turned and disappeared into the hole.

Bollocks.

All was not lost, though. Dick found nestling in the coral a puffer fish. Upon seeing us, and feeling threatened, it blew itself up to the size of a football. His picture hangs alongside the moray in the bar. There is also a picture Dick took of me. I took one of him also and, when we got back home, we mailed them to each other.

The next day was my last dive with Dick. He was going home. We took more pictures and, as we were swimming along the reef, we heard in the far distance what was obviously an explosion. A few minutes later there was a slight sensation in the water, like shock waves. Back on the boats, the others had felt it too. Mr Pastry came to the same conclusion as us.

The Cubana Airways plane!

And, as if to convince us, as we were in the shallows we found two bundles of tied-up, sodden letters. We collected them and Pastry said he would hand them in to the crash investigators in Bridgetown.

It was Dick's last night. He asked if we would join him for dinner at his hotel. It was a good few miles away. Apparently, he had made no friends there and his wife was lonely for some female company. He couldn't get along with any of the other guests. As he told me,

"Des, they're just too far up their own asses for me."

He offered to pick us up from the bungalow. I said that was great, as long as we got a taxi back. There was no way I was spoiling his night by having to drive us back.

It was agreed.

We went back home and had a few hours' kip to be fresh for the evening. Dressed smart but casual, we waited for Dick in the pool bar with a Pina Colada.

He picked us up on time at seven thirty. The place he was staying at looked really expensive. Sandy Lane it was called. I don't know about Dick and his wife, but we would definitely have been like flying fish out of water.

Flying fish? Get it? Never mind.

But what a fabulous place!

I should have been a doctor.

Dick's wife was beautiful. So elegant. She oozed sophistication but was just so down-to-earth and ordinary. Sadly, I can't remember her name. She got on with Teresa like a house on fire. They never stopped talking all night. I'm sure Teresa was starved of female company as much as she was.

It was great. It left me and Dick to our man talk. He recommended the New York steak and I left the wine choice to him. To be fair, I hadn't yet been introduced to fine wine, preferring beer, which we had also.

The steaks arrived with a little plastic bull stuck in them, designating rare, medium or well-done. Pretty posh. Being a butcher, I knew my steak and this was the best. A second bottle of wine landed from somewhere. After that, without any asking it seemed, came four Calypso coffees, piping-hot with cold cream, just how I like them.

After dinner, it was cocktails by the pool until 1am.

As much as I insisted, Dick would not let me split the bill. Which I don't like, by the way. He and his wife said that we had saved their holiday, and it would all go down as a tax break anyway.

After what can only be called an emotional farewell, our pre-booked taxi was waiting to take us home.

• • • • • • • • • •

The next dive was to turn out to be my last. It was unscheduled but unavoidable.

I was given a new partner.

Sharon. A Canadian. She was tall and slim, with a perfect figure and suntan to match. Have you ever seen the film *Dr No?* Honey Rider walks out of the sea with a large conch seashell. That was Sharon.

Les asked me to keep an eye on her. She had limited experience and I was now his senior diver. I agreed reluctantly. I thought her white, transparent bikini was a bit inappropriate, but I didn't think it was my place to raise the matter.

On the beach I briefed her on the dive, saying truthfully that Les (Mister Pastry) had put her in my care as the senior diver.

"It's best," I told her, "That you swim ahead of me so that I have you in my sight at all times."

She was a friendly girl.

"Sure, no problem Des. I'll feel a lot safer knowing you are right behind me."

I wasn't sure if she was taking the piss. But it seemed as though we were going to have the crack.

In the water, everything was going to plan. With Sharon comfortably finning just ahead of me, I got off a few shots. Not my usual subject, but interesting enough.

To be fair to Sharon, it was her who spotted it. An unusual fish. It was like a huge, blown-up head with a tapering body. It had huge, fat lips and a friendly, happy face. I took a few frames quickly in case it swam away. However, it was so friendly it came up to us and posed. It reminded me of some cartoon fish I had seen somewhere.

We were getting low on air and I didn't want to risk any more dive time, because of Sharon. I signalled her back to the boat. I waited behind her as she removed her tanks and passed them to one of the divers.

With her hands flat on the side of the boat, she heaved herself up. To my horror, the water dragged down her pants to her knees. She was struggling about trying to get in to the boat. If she'd been a bloke, I'd have given her a hand up. I didn't know if I should try and pull her pants up for her or not. I decided against. Just in case. She struggled into the boat and sorted herself out.

Damn. I was out of film.

On the boat, I complained to Les that my ears were hurting. He seemed concerned and, after asking a few questions, he looked in them.

"No more diving for you, Des," he told me with a certain finality. "You've got a fungus infection."

He told me what it was, but I can't remember the name. He advised me to call in the pharmacy on the way home and get some drops. He said I had to have them right away, or I wouldn't be flying home.

"And get some strong pain-killers, too," he added. "You might need them."

At the *Coral Reef,* we had a few beers but didn't hang around. We went home via the pharmacy.

The chemist took a proper look and confirmed Mister Pastry's diagnosis. No more diving. It didn't really matter. Tomorrow was the last day anyway. I decided then that we would get up early and go island-seeing for the day.

Back home, I cooked dinner and we sat on the balcony for our evening chat. We hadn't talked a lot, other than about my ears. With drops and pain-killers inside me, we relaxed with our bottles.

"How was the dive today, then?" Teresa asked. "Were you all right with Sharon? She seemed nice."

"She was, honestly, yeah," I said. "It was a good crack all round." "Oh good."

"Yeah. She spotted this fish with great big lips," I said casually. "A bit like her, then," Teresa grinned.

"Yeah. I tried to get a photo, but I ran out of film." "Oh, what a shame," she answered sympathetically.

"It was," I said. "I'd never seen anything like it before."

Early, after breakfast, we loaded up with some water and a packed lunch and, armed with a map of the island, went exploring. Away from the shanty town, the place wasn't so bad. It being very early, I noticed that, in the populated areas, there was great activity in the sea. All the locals were taking an early dip.

I was later to learn that many of them had no proper toilets and used the ocean as a WC. Nice! The further away from people we got, the more beautiful the island got. A bit more like I was expecting. The place we were heading for was a well-known cave site. Sorry, I can't remember the name. We found them after a couple of hours. I was glad we did. It was getting really hot. There wasn't a cloud in the sky and the midday sun was relentless.

Driving in the open Mini Moke with the breeze cooling you, it wasn't so noticeable but, in just shorts and T-shirts, I realised that, even though we were reasonably well-tanned, we were starting to burn.

The caves were huge, black chambers with walls of white crystal nuggets that glowed in the dark. You walked in from the rear off the sandy beach and, ahead of you, at the open mouth of the chamber, was the sea. It was quite rough and the rocks outside foamed white. Walking carefully from rock to rock outside the caves, the sea looked deep and fearsome.

The spray was light and refreshing. We stood on the rocks in the sun cooling off, rubbing ourselves down.

We decided on lunch in the shade of the caves on the beach and walked back through to the beach and scrub sand. The cool-box had kept things nice. I let Teresa have the water, and I had a couple of Red Stripe beers.

We summed the caves up. It was nice to have seen them, but a cave was a cave. I voted to turn back towards the bungalow and have a beer at the *Coral Reef Club*. No arguments there.

The divers were well into it when we got there. We tried catch-up. Mr Pastry gave me my certificate for the twelve dives done. It also hangs on the wall of my bar. Sharon told me how much she would miss me and, with the usual handshakes and kisses, we drove back home.

It was the last night. We opted for a meal in the pool restaurant and had the flying fish again. Beer, vodka and cocktails later, we paid the bill and said goodnight to the staff.

On the balcony that night, we had our last drinks in Barbados. Mid-morning the next day the coach collected us and, after an uneventful drive to the airport, we were in the air and on our way home via the Azores.

The French Concorde was on the ground there, waiting to take off. What a magnificent sight. It had only just come into service in 1976. When I heard about the French Concorde crash a few years ago, it crossed my mind that perhaps it was the one I saw. After all, there weren't that many of them.

The flight was not as much fun as the outward journey and, on landing in England to a cold, wet, miserable day, I knew I was back in the real world.

The realisation hit me. Hit hard. We were getting on towards Christmas, the busiest time of my life, and I had no car and no driving licence.

Twenty-four

Christmas was difficult, but I'd got through worse things in my life. Things were normal at home. I was not going out at night, just not getting in until later. Teresa was picking me up from work and we spent a lot of time in *The Norton*.

Sundays, she took me diving to meet Terry at Stoney Cove. Brian had disappeared off the scene, as it was hard enough sorting me out without the added complication of him.

Easter was coming and the club was organising a week at Fort Bovisand, a large diving centre down south. I was supposed to be going with Terry who, by the way, was the one looking after and washing my diving gear.

Instead of Fort Bovisand, I stayed a few days at the flat with Teresa.

It was coming around to that time when I would be getting my licence back. July 1977, about a couple of months away. It was a long twelve months and made me wonder what it must be like banged-up in prison. At least I was free, and still going to the pub every day. I made my mind up there and then - I was going to try and never go to prison.

I started to think about a car.

I was going to steer clear of anything expensive and exotic. You stick out like a sore thumb to the coppers. I decided on something small that didn't draw too much attention to you.

I scoured the used car columns and found just what I wanted. A beautiful, almost-new citron-yellow MGB GT. One owner, low miles, spotless. The man selling it was about my age and, when I told him I couldn't drive at the moment, he brought it around to my house.

I instantly fell in love with it. He was selling it, he told me, to buy a brand-new TR7.

Good luck with that, I thought. I loathed the TR7. It looked a crock of crap to me, but I told him how lucky he was and, when I could afford one, I'd buy his if he was selling it.

I paid him cash there and then for the MG, and his wife, who was following him in another car, drove him away.

Now I swear to God, the next thing is the gospel truth. About two months later, before I even had a chance to drive it, he was knocking on my door asking if he could buy it back. The TR7 was the biggest junk-box he had ever driven. I could have told him that. Sympathetically, I assured him he'd get used to it, and all the difficulties he was having were only teething problems. Some hopes.

July arrived and I was a born-again man. But I was making changes to my life. A year without my licence had taught me a valuable lesson. I didn't need a car to get to the pub. I needed a car to get to work and conduct my business deals and just generally get about any time I needed to.

Evenings I would drop the car off home. I'd tell Rita I was getting the bus at the top of the road to *The Norton* then walk to the stop just around the corner, where Teresa would pick me up. Sundays stayed mostly the same, Teresa taking me to Stoney Cove.

Late that September the club had organised a week's diving in Oban, Scotland. Naturally, I couldn't miss that and Terry was really anxious to go. He'd never dived with anyone else and wouldn't go if I didn't go.

Well that was the story, anyway. I phoned Terry and told him I wouldn't be at Stoney that weekend and under no circumstances phone the house.

"Roger that," he laughed. "See you next week."

First thing Sunday morning I was round the flat, breakfast was ready and all the dive gear stacked by the front door, ready for loading. The car was full of petrol and raring to go. The route was all planned out and we set off in great spirits.

The car was a dream to drive. It had overdrive and, if you asked, it would do ninety with ease.

It had an eight-track stereo and we had a stock of tapes to play. We only ever played three. Smokies greatest hits, Glen Campbell's greatest, and the Moody Blues, whom Teresa loved and who were close personal friends. The car ate the miles and, the further north you got, the less traffic there was.

When we crossed the border into Scotland, the traffic was practically non-existent. The club members were staying in chalet-type accommodation a few miles outside Oban. We were booked into the

Manor House Hotel, right by the water in the town of Oban itself.

A grand place it was, too. Small and intimate. Huge, open fireplaces in the lounge and dining room. There were crisp, white tablecloths with candelabra and flickering candles. The room was sumptuous and spotless, with a four-poster bed and its own small fire.

I knew then I wouldn't be doing much diving. The food was possibly the best I'd ever eaten up until then. It was certainly the most expensive. After dinner the lounge, with its small, intimate bar, was deserted. It was out of season and would be quiet most of the winter. The deep, leather sofas were so comfortable you didn't want to get out of them.

But it had been a tiring day and it was time for bed. In the morning, I would find the club members and tell them where they stood.

They were dossing around on bunk beds in a couple of chalets in the woods by the coast about five miles north of us, when we drove into the gravel area out front.

Tony Dean, the main man, was the first to greet us, with his ever-present wisely grin. "Des! How good of you to join us," he greeted us jovially, "And the lovely-as-ever

Teresa."

"Hello Tony."

"And are you comfortable in the *Manor House?* I've been there, it's a lovely hotel," he went on.

"Really lovely, Tony," Teresa smiled. "So grand and olde-worldly."

I knew I was in for a ribbing, and I hadn't even seen the lads yet.

"When will you be joining us?" Tony asked. "The diving is good. Of course, nowhere near as good as Barbados."

I chuckled. I just had to take it on the chin. "A tad colder, I reckon," I said.

"Just a bit, I think," he smirked. "The lads are in Nos 7 and 8 if you want to say hello." "I will," I said. "There's just one thing; I'm not going to be doing much diving." "Really? But you've come all this way!"

Teresa was laughing. "Why?" he asked.

"Tone, give us a break, will ya?" I moaned. "We're booked into the hotel for the week, as you well know, but there is one dive that I really want to do."

Interrupting me, he said, holding his hand in the air as if he'd had some great vision, "The *Breader*."

"You got it!" I exclaimed.

"Well how did I know that?" he asked, in a deliberate tone. "What about it?" I said seriously.

Suddenly, knowing the joke had run out of steam, he said, "Sure thing, Des, we're diving her Wednesday. Estimating the tide and conditions, it'll be a ten o'clock dive. Be here at nine for the dive team briefing. We're boarding the boat at nine thirty and should be over the wreck by then. Got all that?"

"Sure," I said, "And thanks."

With that, we caught up with the lads and, over some tea and coffee, I suffered the relentless expected piss-take, ending with,

"So, when's the best night for us all to come and have a drink at your place?" "See ya," I told them, turning my back and raising the middle finger.

"See you, Teresa," someone called. "If he gets too boring for you, you know where we are."

"Sure do!" she waved.

The *Breader* was a transport ship sunk by a German sub in the war. It was carrying Army supplies and was standing upright on the bottom, still with Land Rovers chained to the deck. I'd heard a lot about it from divers who had been on it before. It was beginning to deteriorate and wouldn't be at its best for much longer. It was the sole purpose of my trip.

Wednesday Teresa dropped me at the Dive Centre and said she was going to explore the area. We agreed to meet back here at midday.

I attended the briefing and got kitted up in the large, communal changing-room and showers. Only tanks and flippers to add. All set, about eight or ten of us boarded a quite sturdy and large dive-boat.

I teamed up with one of the guys from the club. It was a short trip to the *Breader*, and the boat anchored directly on top of it. The deck of the ship was about 100ft below us, I was told. Just follow the anchor line down.

The water was murky, with vis 30ft at the most. Thick strands of kelp grew from the seabed to way above the deck of the ship. The Land Rovers were there, chained on both sides, but were starting to crumble and disintegrate. A few had fallen into the wreck where the deck had collapsed. The whole ship was deteriorating rapidly. It was completely overgrown and encrusted in sea life.

Fish were living in it everywhere. Exploring along it, though, was fascinating, trying to imagine how it must have been for the crew. Men being blasted and jumping into the water for their lives. We had only twenty minutes air and bottom time, and that was soon up.

Swimming back to the anchor line, I saw a massive crab scuttle from its hiding-place and across the deck to another. Triumphantly, I swooped and snatched it up. Waving it at my buddy, I started climbing the anchor line, the crab's legs dangling below my hand. Suddenly, without warning, the little bastard managed to sink a claw straight through my glove and into my finger.

Air gushing from my regulator and blood seeping from my hand, I howled with pain and shook the little bleeder free, to watch it slowly sink back to its home on the *Breader*. I'd seen the crab as a great dinner. He obviously wasn't ready for the feast.

Of course, it made everyone else's day and, by God, was I glad I wasn't going to be there that night.

I was showered and changed. Teresa picked me up and took me back to the hotel. I told her all about it, to her great amusement, over afternoon drinks and dinner.

It had been an unforgettable dive all the same and, after a few more relaxing days of pubs and sightseeing, it was time for the long drive back with Glen, Smokie and The Moodys.

Happy days.

Twenty-five

Unbelievably it was nearly Christmas again.

The strain of living two lives was beginning to tell. It was beginning to tell on Teresa, too.

Her family were beginning to ask questions. Where was Des again at Christmas?

She was beginning to feel the pinch financially, too. The money she'd had when she left Bill was gone. The job she had part-time at the flats was gone. They were all sold. The company did, however, keep her on with less hours. But that meant she had to travel to the outskirts, where the new building was going on.

I was having to help her out with money.

Her divorce was grinding slowly. Apparently, Sabbath were broke, despite earning millions in America and worldwide. She was being asked to negotiate with Albert Chapman who was, apparently, their Road Manager.

Teresa knew him, of course, and indignantly blustered,

"If he thinks I'm negotiating my divorce with the bloody Road Manager he can think again."

New Year 1978 came and went. Sid Pickering was booking our annual holiday in June to Santa Ponsa. I don't think Teresa was too happy to hear about that. I sensed that now, understandably, she was wanting more.

Well, I'd made it plain from the beginning that I had no intention of breaking up my happy family. Apart from everything else, there were a lot of very complicated finances in play. Even that aside, there was my main priority, Angela. There was no way I was wrecking her life with a messy divorce.

As much as I loved Teresa, I had to start thinking on my feet. I honestly didn't want to hurt her. She had never done anything other than love and worship me. I wasn't going to do anything to hurt her if I could possibly avoid it.

She deserved that, at least.

I'd never even thought about us breaking up before. However, when you look at it logically, how could it ever have gone on for ever?

There was inevitably going to be pain involved. On both sides. How to keep it to the minimum was the problem. I hadn't stopped loving her. That was the big problem. The other big problem was that I also loved my wife and daughter.

At the end of the day, it was my mess and I had to sort it out.

Things carried on as normal until Santa Ponsa came around. I was looking forward to my two weeks' diving with Raymond and Danny. Teresa was upset and I can understand that, but she knew the score and, besides, I needed a couple of weeks' break.

Angela was ten now and we were having fun, as normal families do. We swam in the pool, played on the beach and had a few rounds of Crazy golf. We had lunch at Jaimi's, dinner in the hotel and nights out, usually at *Johnny Valentino's*.

I discussed my situation with the lads, who all knew her, and I received drunken advice of all sorts. None of it really helped.

Sid Pickering, being a bookmaker, loved his horses and always missed a little bet on holidays. One day, one of us came up with the idea that I pick a horse and have a bet from the daily paper when it got to the hotel about midday.

The papers were always a day behind in those days in Spain. Over a beer, I picked a horse and bet 1000pts on it, at the price it was in the paper. By the way, 1000pts was about five quid then.

The next day in the bar, when the papers arrived, Siddy checked the racing results and jubilantly relieved me of 1000pts. Not good! Sid wanted me to have another bet for tomorrow but, not normally being a betting man, I told him maybe later. I had a plan.

Public phones were cheap to use. There was one not too far from the bar. I said to Sid,

"I'll be back in a bit," and, with a pocket full of change in my shorts, I made a call to Uncle

Bill in England. Bill liked a bet and knew his horses. He had a bet most days. He was surprised to hear from me.

"Hello Bill," I said, "I've got to be quick, I'm in Spain."

"I know," he answered. "What do you want?"

"I want a winner on the horses," I told him.

"Don't we all," he said, "But the racing's nearly finished."

"I know, that's why I'm ringing. I want a horse that's won at about 10 to one today. Any horse will do." I explained to him why. I can't remember the horse, so we'll call it Fred Blogs.

"Fred Blogs won the three thirty at Ascot," he told me.

"Great," I said. "Now, to save me time and money, I'm going to call again tomorrow. Can you have a winner ready for me? Something around that price."

"No problem." "Thanks."

I hung up and wandered back to the bar. I ordered San Miguel Le Grande.

"Let's have a look at the paper then," I said to Sid. "I'll try and get my money back."

With a big grin Sid passed over the paper. "Good luck."

"I'll probably need it, too," I grunted.

I cast my eye over the racing pages.

"I haven't really got a clue," I said. "I think I'll just go with the tipsters."

"Good luck with that," Sid said scathingly. "You'd be better with a needle." "Oh well," I sighed, "Fred Blogs, three thirty at Ascot."

Sid checked it out but didn't seem too worried. "Not what I'd have picked, but you never know."

I do, I thought.

The next day, the papers came to the hotel and Sid collected one from Reception. I was sitting outside the pool bar with a pint.

"Papers are here," he called, waving it in the air and coming over to sit with me. "Let's check the results."

"Blimey," I said. "I'd almost forgotten about our bet. I'd better get my money out. What was it called? Fred Blogs?"

Sid was fumbling through, trying to find the results.

"I don't friggin' believe it," he cried. "Fred Blogs. Ten to one, first."

"Well, bloody hell, thank God for that," I sounded relieved. "I was pretty sure I'd lost another 1000. So, does that mean you owe me 11,000pts?"

"Yes," he sulked. "I'll have to change a traveller's cheque. I'll pay you later." "OK."

I phoned Bill a bit later for another winner. And, after another win the next day, Sid announced that he couldn't afford any more bets, explaining that the concept of horse-race betting relied on the other losing punters covering the winning bet. As I was the only person betting, he had no chance of recovering his losses.

I fully understood and, as I felt guilty anyway, was relieved to end the game.

With the holiday over, I found that I was not, on this occasion, looking forward to going home to Teresa. The clouds were gathering. I felt doom and gloom in the air.

The End is Nigh.

Twenty-six

I didn't think so at the time, but it seemed that fate was to lend me a hand. Rita was pregnant.

It was, without doubt, a major complication. We discussed it, with differing opinions.

Angela was ten and Rita didn't want to start all over again. I was morally against abortion, but had always conceded that, if it were done immediately, before the baby was formed, then I could live with that.

So, it was decided.

An appointment was made at a private hospital in Edgbaston, Birmingham. It was a Saturday; a good mate of mine, Jeff, who worked for a large butchers' chain, Walter Smith, took the day off to do the shop.

I went with Rita and she had the operation. She stayed in overnight and I picked her up the next day.

Obviously, all arrangements with Teresa were out the window. She was hassling me, wanting to know why and all the details. I told her Rita had been taken ill and was in hospital. Where? I told her. How was she? She was all right. What was wrong with her? "They're not sure."

It went on and on.

"For Christ's sake," I exploded over the phone, "I don't need this." She was starting to figure it out.

The Priory, Edgbaston, a well-known abortion clinic. When I finally got around to see her at the flat, she was sobbing her heart out.

It was pitiful to watch.

She was curled up on the settee, just inconsolable. I tried to comfort her. I didn't try to lie.

She knew. How could she have been so stupid? She'd always convinced herself that I would never have gone with Rita even though she was my wife. What a fool she was.

There was nothing I could do or say to console her. I genuinely thought that she may commit suicide. I spent every minute I could with her. Every time I got to the flat she was still on the settee, sobbing. She wasn't eating, drinking or sleeping. Just lying there in the dark with no TV, nothing.

I was worried sick. I just didn't know what to do. I made her food and did her drinks. Eventually, I seemed to be getting somewhere. I sat beside her and held her in my arms.

We talked.

For the first time.

I told her how sorry I was. We discussed it.

In her heart of hearts, she had always known that it could not have gone on for ever.

We both agreed that we loved each other and that should never be forgotten. There was no reason to ever hate each other. We wanted always to remain friends, even though in spirit only. We wanted no acrimony.

There was too much love in us both for that. It had all been good.

It had never been bad.

And, by God, it would stay that way. We kissed. We kissed again. And then we made love.

For the last time.

We made plans for our parting. We wanted to end on a high note. I suggested that I pick her up on Friday evening and that we could go to *The Wilkin* in Brownhills. Where we had our first date.

We did that. Over a nice meal we held hands and wished each other luck for the future.

She believed it was only months away to finally getting her divorce sorted. We discussed the flat. I told her it was hers. I wanted nothing from it. She said she would sell it, and planned to visit her sister, who lived in Canada.

I told her I had no plans. Without her in my life, I only had Rita and Angela. Eddy was older and I wanted to share time with him. It had been a great few years and I never regretted a minute of it.

We laughed at old memories. We got melancholy over some.

But, best of all, we had no bad memories. Just sad memories.

And thoughts of the inevitable, lonely times ahead for us.

I drove her back to the flat and parked in the car park at the back.

I looked into her big, sad, green eyes for the last time and kissed her goodbye. She cried and got out of the car, not looking back.

I sobbed my heart out all the way home.

PART TWO

Twenty-seven

I was gutted.

I'd lost the love of my life, my best friend and soulmate all in one go. Why?

We both still loved each other. We hadn't had a row and fallen out.

Neither of us had fallen in love with someone else. But we both knew it had to end. She was on my mind twenty-four seven.

I was sick with worry for her over her state of mind, health and safety. I couldn't phone her, that was part of the deal. It would make things worse. We just had to try and deal with it and hope that time would heal.

I did the only thing I knew how to do. The same way I dealt with all my pain. I shut it out of my mind. It was like a chapter in a book. When it ended, you turned over the page and started a new one.

I concentrated on work. Working hard at selling meat. I bought another shop on the Stratford Road in Sparkhill, a good Irish area. Big meat-eaters. I sold one of the houses for a good profit and bought a big house with about twenty flats and bed-sits in it.

I went in half shares with Uncle Bill. We didn't keep it long. It was a lot of hassle collecting the rents. A housing association offered us a double-your-money offer if we could give it to them with vacant possession.

We gave notice to all the tenants but some didn't want to go. Eventually, we were down to the last few. Offering incentives got rid of most, but there were two who dug their heels in.

I knew a couple of persuasive guys who made them see reason, and the place was boarded up.

One Saturday in November, I was drinking as usual in The *Buccaneer* with Rita and friends. One of them was a bloke about my age, who worked at the airport. He had a couple of free tickets to go to Tel Aviv, in Israel, somewhere he had always wanted to go. The person he was going with had changed their mind and said they weren't going.

Roy Bridgewater didn't want to go on his own. Sensing an opportunity,

I said I would go. "Great," he said. "We go on the Sunday in two weeks' time."

That gave me time to sort out the shops. Also, I thought it would be just the distraction I needed.

We were flying El Al, the Israeli state airline. The security was unbelievable. We were interviewed individually by two people. We were asked numerous questions. It was an interrogation, really.

We were obviously cleared to go. We were going on a Jumbo jet. I don't think there were more than twenty people on it, unless they were all in first class. We were in the middle section of the plane, with a huge TV screen in front on the wall.

We had the whole area to ourselves, except for three Mossad Israeli agents watching every move we made. With all that room, we raised all the armrests and made beds. It was going to be a long, comfortable flight.

There were loads of food and drink and, after a film, we settled down for a sleep.

In Tel Aviv, we passed through security and boarded a coach for the *Park Hotel* in Netanya.

The hotel, what I can remember of it, was nice. It was all buffet food, so we helped ourselves to whatever looked all right. The food was a little strange.

As it was our first night, we decided to find a bar not too far from the hotel. I thought this was a country with troubles and it was best to play things safe. Wandering around the streets were a lot of military personnel. As many were women as men. Apparently, the women also have to do national service.

The bar we found was pretty quiet. The natives seemed friendly and all spoke good English. We sat at the bar, chatting and finding out about local customs. They seemed as interested in us as we were in them.

We told them that we had booked tours with a local company so that we had somewhere to go every day. They asked us where and we told them.

Jerusalem. They agreed it was fascinating and a must-see. The Crusader City. Another wonderful, historical site.

A guided tour of Tel Aviv. A good idea to go with an organised tour. So far, so good, I thought.

Eilat, on the Red Sea. I wanted to dive the Red Sea. Fabulous, you'll love it, was the general feeling.

The Golan Heights. Sharp intake of breath. Not such a good idea, it seemed. There was still regular shelling up there by the Palestinians.

Masada in the desert by the Dead Sea. Apparently, a trip not to be missed. One of the great stories of the New Testament. Definitely going to love that.

As I've said, the natives were really friendly. As we sat on our bar stools a beautiful, dark-skinned Israeli girl with shoulder-length, black, silky hair, smelling faintly of lemon, slid onto the stool next to me.

Anxious as always to improve my foreign relations with the local population, I asked her to have a drink. She happily accepted and we introduced ourselves to each other. Des, Roy and

I'm sorry, I can't remember her name. Anyway, there were far more memorable things about her than her name.

We chatted away happily. She was very interesting and also wanted to know all about life in Great Britain. To my surprise, she bought a round of drinks. We learned later that, that was perfectly normal.

At closing time, she kissed us both on both cheeks and said goodnight.

We had to be up really early in the morning. Five o'clock. Roy was not used to it and moaned loudly. It didn't bother me, I was used to it.

The breakfast was all laid out. Cold meats and cheeses and breads. It was OK, I suppose, but you can't expect bacon and sausage in Israel. The coach was early and had many pick-ups to make for the tour.

Today was Jerusalem.

I have to say that, to this day, it has to be one of the most exciting things I've ever done in my life. We visited the Mount of Olives and Calvary, where Jesus was crucified. We went inside The Holy Sepulchre. The tomb of Christ. You had to duck to go into the cave and kneel by the slab of rock his body lay on. There were nuns overseeing things, and a voluntary collection dish. As I ducked out, I couldn't help feeling I was having an out-of-body experience.

Whether you believe in God or not, something happens to you in there.

Next, we did the Via De La Rosa. The Stations of the Cross, as we know it. To walk on the very same streets He did two thousand years ago is incredibly awesome and moving.

Finally, we went to the Wailing Wall and watched as Jews all lined up to pray and meditate before it, wailing loudly.

A never-to-be-forgotten day.

Full of excitement, we went back to our bar from the night before and shared our day with the eager ears of the bar staff, who seemed so happy that we were enjoying their country and respecting their culture.

The girl from the night before appeared like an angel by my side.

"Hello again," her silky voice said, "How was your day? You must tell me all about it."

We did, enthusiastically. She listened intently and asked us questions as if it were a school exam. The smell of lemon hair and intoxicating perfume drifted onto me as she fixed me with large, dark-brown eyes. Her long, brown, smooth legs shined as if lightly oiled.

She was bordering on exotic and erotic. There was something in the air. I could feel it.

She leaned across and whispered in my ear, "Come with me. I know a hotel. I pay. You come with me."

My heart was beating fast. It was a proposition that no man in his right mind was going to turn down. But I had my fears. It seemed too good to be true. And you know what they say.

I knew that what I was going to do next could completely blow the whole thing out of the water. I turned to Roy and said,

"Could you look after these?"

I gave him my gold bracelet and wallet. She looked at me, hurt and offended. "You don't trust me," she said.

"I'm sorry," I said, "But you must understand. I'm in a strange country, alone, and I have to be cautious."

I was thinking that I could go outside and be beaten and robbed. If she still wanted me to go with her then at least she would know there was nothing to rob.

There was no one more surprised than me when she took me by the hand and led me a few streets away to a small hotel, where she paid cash and took me upstairs.

It was barely light in the morning about five o'clock. I found my way back to my room at the *Park Hotel* and showered and changed for our early-morning pick-up for The Crusader City.

Roy did nothing but moan. He was bloody tired. Why did they have to pick us up so early? "Well at least you've had some sleep," I told him.

"Yeah!" was all he said. No sense of humour.

The city was interesting. Medieval, secret passageways. Walls, dungeons, defences. All the usual old castle stuff. It was worth seeing, but not a patch on the day before. Nothing could be a patch on the day before. Who was that girl, and why me?

I kept going over it in my mind. Doing replays. Definitely the most exciting out-of-mind experience I had ever had in my life.

Not to mention Jerusalem!

A Pierce Brosnan smile crept over my face. It was all very James Bond.

My mind drifted to this evening coming. I wondered if it could be a case of *You Only Live Twice!*

I hoped so.

But there was no sign of her.

You could almost hear me calling, "But it's my turn to pay!"

• • • • • • • • • •

Jesus Christ. Talk about the Wailing Wall. Roy Bridgewater - moaner extraordinaire.

It was Golan Heights day. Another early start.

The tour guide gave us a full account of the Palestinian war. I found it fascinating. I'd always been keenly interested in all things Israeli. They took no shit from anyone. I admired that. You could see from the top of the Golan Heights why they were never going to give it back.

They had won it with blood. It controlled the whole expanse of the valley below it.

Whoever wanted to take it back would have to be prepared to pay a heavy price. It was a quiet day today, I was told. No shelling. The previous day they had been shelled and had to take cover.

Inspired by my history lesson, we made our way back to Netanya. We were early and went to our bar for a few drinks before dinner.

We had some sort of hot food and went out again to our bar.

My girl was there for catch-up. Roy excused himself, saying he could do with an early night.

"Do you mind?" he asked.

"I'll be OK, mate," I reassured him. "You carry on. We've got an early start tomorrow."

It was Masada and the Dead Sea. "See you later."

Head to head, we talked all night. She was great company and very easy on the eye. The night ended as I had hoped it would.

My treat this time.

· · · · · · · · · ·

Masada was amazing. I had read the story, but standing on the walls of the city in the middle of the desert was biblical and awesome.

The Romans had laid siege to the city. And despite all efforts to penetrate the walls, they could make no headway. The defenders were self-sufficient within the walls. They had ample water from deep underground sources. They had livestock enough to breed and grew food enough to hold out indefinitely.

The Romans, however, were not to be outdone. Their engineers came up with an ingenious plan that, simply put, would be a road to the top of the wall that would enable them to attack the city and take it.

I haven't got time to spend on all the details. In short, they managed to break the siege and enter the city. However, upon its capture, they discovered that every man, woman and child had either committed suicide or been put to the sword if they needed help with killing themselves.

Nobody was prepared to be a slave to the Romans and spend the rest of their lives working in chains and the children forced into fighting for the Legions.

Death was better, in this case, than life.

After a couple of hours' freedom to explore the city, we were taken to the Dead Sea. The water is so thick there with salt, they say it is impossible for a man to sink. Also, it is said to have many healing qualities.

I came prepared to swim. I had my trunks on under my shorts, and a towel. I wandered in up to my waist and sat on top of the surface, bobbing up and down. It was true what they said - you couldn't sink.

There were fresh-water taps on the beach to wash down with. It is very important to wash off the salt as the sun is so fierce that you could burn seriously if you didn't.

Roy didn't swim. He stayed on the coach, sleeping. He was tired. It was the last tour tomorrow. The Red Sea. For that we had to go to the airport and catch an internal flight to Eilat. I didn't know how he was going to cope with that.

Back in Netanya we were straight to the room to shower for dinner. It was cold chicken, fish, cheese and breads. We ate quickly, sinking a bottle of cheap wine, and set off to our bar for beer. Obviously, I was hoping that my "bird", as Roy called her, would be there. He was hoping so, too. It meant he didn't have to stay out with me and could get back to the room and have a few fingers of Scotch.

Neither of us was disappointed. About nine thirty my "bird" breezed in, smiling. She plonked herself down on the stool next to me and gave me a smacker on the lips.

Wow. She was dressed to thrill. White linen shirt with two pockets on the front and a short, denim skirt and sandals. It was simple but, with no bra, very provocative. She had a sprinkling of jewellery and lemon and perfume, hair and skin.

Heads turned.

Roy finished his beer and said he was going for a few in the hotel bar and an early night.

I never really figured out why she had come on to me that first time, and I could only be thankful that she had. I'd thought, wrongly, that she had wanted something from me.

"Tonight will be the last time," she told me. "This time we go Dutch? Is that how you say it?

We share the price, yes?"

"I still have another night," I told her.

"I do not," she told me back. "Tomorrow I must go and rejoin my unit. Maybe to fight.

We will never meet each other, ever again." Somehow, I think that explained it all. Another goodbye.

• • • • • • • • • •

Roy was staying in bed. He said he'd done enough travelling and wanted a quiet day in the sun on Netanya beach. I wished him a nice day and was picked up and driven to the airport.

The plane was a domestic flight. It was filthy and deep in litter. You could kick the fag packets and sandwich wrappers aside to get in your ripped and dirty seat. It was packed and was obviously just a glorified bus.

I was glad to get off in Eilat. The temperature was a lovely 70°F, not a cloud in the sky. Eilat was just in its infancy then, with only a small strip of hotels and bars.

I found what I was looking for.

Lucky Divers. A decent-size shed on the beach, with an awning on the side with dive tanks and equipment stored beneath it.

Eilat was, literally, a tiny town on the shores of the Red Sea with dozens of miles of desert behind it.

I'd been booked for a dive that day with three others. The instructor was French and, with not much ceremony, we collected our tanks and paddled out in the shallow, warm water to a quite substantial inflatable with a fifty-horse Mercury outboard.

Fins, mask, snorkel and tanks.

That's all you needed. We cut through the water at speed out to the dive site. With the inflatable anchored, all five of us slipped over the side and followed the instructor. It was, of course, amazing.

The water was teeming with fish. The coral was alive and vibrant. The visibility was easily a hundred and fifty feet. On its periphery could be seen the shadows of larger predators, staying warily away from us divers intruding on their territory.

The depth was only sixty feet and, with the water so warm, our air comfortably lasted the expected sixty minutes. Back on the inflatable, we all agreed it had been worth all the effort to get there. One dive, however, was nothing near enough, but I had accomplished what I set out to do.

I wanted to say I'd dived the Red Sea.

The plane seemed even dirtier on the flight back, and I'm sure it was. My thoughts, though, were of a great dive and our last night in Israel.

When I got back to the hotel, Roy was waiting in the bar for me. I asked if he'd enjoyed his chill-out day on the beach and he started laughing uncontrollably. When he finally got himself together, he told me his story.

Apparently, it had been huge news on the TV and radio all day.

Terrorists had taken over the school in Netanya and were holding the children hostage. The Army was obviously called in and, while sunbathing on the beach in the dunes, Roy found himself in the middle of the operation to capture back the school.

He told me he had nearly been run over by an armoured car racing along the beach.

Several troop-carrying vehicles had sped past him armed to the teeth.

According to the news, the troops had surrounded the school and attacked, killing all of the terrorists.

The children were all safely freed. We both had a good laugh about his quiet day on the beach and, after dinner, hit the bar. Big time.

Somehow, though, the place didn't seem quite the same without my "bird".

Twenty-eight

It was almost Christmas. I was determined to have a good one. There was a glut of turkeys and plenty to be had cheaply. Myself, the Old Man and Eddy bought together. With our larger buying strength, we could get them even cheaper.

Eddy was seventeen now and could drive. We were meeting up more and more after work for a few beers. We played golf on Wednesday afternoons with Uncle Bill and a friend of his, Tom Blow.

Eddy was a good golfer and I could never beat him. We started playing the new in-game, squash, and I managed to hold my own at that with him, but he was gradually getting the better of me at that, too.

It didn't really bother me that much. I had never really been a competitive player at sport.

For me it was just a way to stay fit while having some fun.

Christmas was, indeed, profitable that year and, with the rents from my flats and houses, I was pretty flush. Also, my overheads were down by a flat in Erdington and several nights out with Teresa.

I was staying in more and enjoying a normal life with Rita and Angela, who would be eleven in March. My military manoeuvres with the Israeli Army had helped me through a bad period.

It was February 1979 and the annual Boat Show at the NEC. I always went and loved drifting from one hall to another, having a pint in every bar along the tour.

A particular stand caught my eye. H&L Boats, T/A Hustler Powerboats. They had half a dozen speedboats of different sizes and specifications.

The one I wanted was a Hustler 146. With a trailer and 50 HP Mercury O/B engine, it had a list price of nearly two thousand five hundred pounds. After some hard negotiating, I ended up paying, with VAT @ 12 %, £1981.60. That was to take the one that was on the stand at the end of the show.

I had argued that, although it was brand-new, it was really second-hand after God knows how many people had climbed into it and sat on it.

It was a spur-of-the-moment purchase. I'd never had a boat in my life, or used one. However, it was yellow and would look great on the back of my yellow MGB GT. Strangely enough, I dived with a pair of Twin fifty tanks that were also yellow and looked great in the back window of the MG.

Of course, I don't want you to think that I was a poser, but the boat looked great parked in the front garden of my house where it was, apart from one excursion, to spend the rest of its life.

Talking about that excursion, it was early summer. After diving at Stoney Cove one Sunday, a few of us were drinking in *The Holywell* when we decided to take the boat on its maiden voyage. Terry knew a place that was ideal. Lulworth Cove, down on the South Coast.

Terry and I organised a date for the trip and prepared for a fairly long drive. He said we could do it there and back easy in a day. We chose a Sunday and made a really early start, at about five in the morning.

I set off in the MG with the boat on the back and picked Terry up from his home in Warwick. We were down at Lulworth Cove by about 9.00. It was very quiet at that time and I reversed the boat down to the water's edge and unhooked it. We pushed the trailer the last few feet down the slip into the water, then just up to the sand while I took the car to a nearby car park.

The diving gear was all in the boat. We got changed, pushed the boat into deeper water and started the engine. Soon we were heading right out. The boat was fast. We found a place well out by a headland that was deep and clear. You could see the bottom. We threw down our anchor and felt it stick and hold.

We started organising our masks, tanks and fins for the dive. I moved towards the front of the boat and felt it dip down. As it did my plastic sandwich box floated past me in about four inches of water. Slightly alarmed, I called over to Terry, who then noticed that the back of the boat had nearly a foot of water in it.

After a brief panic, we retrieved the anchor into the boat and started up the engine. We were about two miles out and aimed the boat as fast as it would go until we were nearly on the shore. We cut the power and aimed at an empty stretch of beach and drove right on to the sand.

The beach by now was quite busy and our rather dramatic beaching created a lot of interest. Soon we were surrounded by a lot of holiday-makers eagerly listening to our shipwreck tale.

It turned out that the boat's drain-hole at the back had not been sealed correctly and was letting in water. We had, had a lucky escape. If we had dived without noticing the water in the boat, we would most likely have passed our boat going down as we were coming up. That would have been fun!

Two miles out!

We drained the boat on the beach and eager hands helped us reunite car, trailer and boat.

Deciding against a few beers in the local pub, we set off back home. That was my first and last voyage of discovery in my new boat. Maybe it was for the best that I never joined the Merchant Navy. For them, anyway!

Twenty-nine

Ed thought it was great when I told him the story in the pub on Monday. In fact, everyone did. I was getting great mileage out of it in the pub and diving club. My street cred was climbing.

It was August 1979. My birthday. I was thirty-one.

"Bloody hell, Des," someone pointed out, "You're on the forties run." He was right.

It seemed to me that I was wasting my life away.

I saw a Jensen for sale. It was one of the very last ones they made before going out of production. It was a beauty, with one owner and low miles. I bought it. I sold Rita's Mini and gave her the MG. She loved it.

I was now the proud owner of an MG, a Jensen and a Ford Anglia van.

It was Saturday night and a hot summer evening. The MG was in the garage for some work. All dressed up, we climbed into the Jensen. It wouldn't start. Would you believe it? The battery was dead.

Nothing else for it, into the van.

When we got to The *Buccaneer*, most people were drinking outside. The Old Man was there with Eddy and Uncle Bill and the rest of the crowd.

Always quick with his wit and sarcasm, Bill quipped,

"Des, is this the height of snobbery? A Jensen Interceptor and an MGB GT in the garage, and you go out in the van?"

Everyone was laughing, including me. I had to admit it was funny.

• • • • • • • • •

Ed and I were like best mates. Golf Wednesday, squash twice a week and the pub several times.

I'd only had the Jensen a week and I was taking him for a run in it to show it off to him. He loved it and was really enthusiastic over it. He had some exciting news of his own for me.

"I'm going to California."

"You what?! When?" I gasped.

"Middle of September," he told me, as if he were going to Jersey. "Who with?" I asked.

"On me own," he told me.

"You can't go to America on your own," I stammered, "You're only eighteen."

"I've got no one to go with," he said flatly. "It's expensive and no one can afford it." "How long are you going for?" I asked.

"Two and half weeks," he said. "I've got it all sorted. Fly from Heathrow to Los Angeles. Hire a car. Drive down to San Diego then back to LA, then over to Las Vegas. A few days there then drive to San Francisco and get a plane to Honolulu for five days, then back to San Francisco and drive down the Pacific Coast Highway to San Jose and all the little towns on the way back to LA for my flight."

"Can I come?" I asked.

"Course you can," he said, "But how you going to swing that? What about Rita?" "Let me worry about that," I told him. "Can you organise things?"

"No problem," he said casually, "The travel agent at the *Wheatsheaf* can do all that. It's only the flights. Everything else is on spec."

"Get on with it, then," I said, "There's not much time."

"Are you sure?" He sounded uncertain. "The flights aren't cheap and I'll have to pay for them cash."

"Don't worry about the money," I told him, "That's no problem."

"I know the money's no problem," he said, a little nervously, "But what if Rita cans it?" "She'll be all right," I assured him; "Leave her to me."

• • • • • • • • • •

I told Rita the truth. And she totally agreed. After what had happened to Phil, there was no way I could let Eddy go off to America on his own.

I'd never forgive myself if anything happened to him. You know what America's like, even the kids carry guns.

I just couldn't let him go on his own. Not at his age. I didn't know how I was going to manage it, but I would have to be there to look out for him. He was the only brother I had left!

It was all true, really. I just embellished the story slightly.

I did have to think about Ed's safety and, besides, I was never going to miss out on a trip of a lifetime like that.

Thirty

I didn't have much time to organise everything but, somehow, I managed to get it all done.

As if there was ever any doubt. At the last moment, the most important part of it all dropped through the letterbox at home.

My American Express Card. I'd applied for one and was relieved to get it. They didn't hand them out to just anyone. But if you can get one, as they say in the ad,

"Don't leave home without it."

I felt a lot happier and safer with it to take. It was a great problem-solver. We were set to go.

The flight, I always think, is a great part of the holiday. We were flying Pan Am, a scheduled flight with a free bar all the way to Los Angeles. Comfortably settled in our seats, we caught the attention of the stewardess.

"Sorry to bother you," I said apologetically, "Could we have a couple of beers when you get a moment?"

"Of course, sir," she replied, very professionally. "Coming up."

We were to find the flight attendants extremely friendly and helpful. The meals were great and, watching the in-flight film, the flight was drifting by pleasantly. We had the girls well-trained after a few hours.

As they would pass down the aisle they would beam. "I know. Two more beers!"

"Thanks."

After a while, they didn't even bother asking. They just brought two when they were coming our way. I must say it was our first experience of drinking American beer. It only came in small bottles and there seemed only a mouthful in it. We soon learned never to buy anything but draught if you could get it.

We landed in LA and took our turn in a long queue at Immigration. They were quite thorough, asking many questions and inspecting my last-minute visa. Satisfied that we posed no threat to America, other than their young girls, of course, they stamped our passports and wished us happy hunting.

The first stop was the bar for a couple of proper beers. Pints. "Bud or Micky?" the immaculate bartender asked.

We looked at each other and shrugged. "What's the difference?"

"I prefer Micky myself, sir," the bartender said, "But I'm sure you'll find your own favourite."

We took his advice and had two Mickies. They were OK.

Thinking of driving, we only had the one beer and sought out the Avis Car Hire desk. "We'd like the cheapest car you've got," I told the girl in the Avis uniform.

"The cheapest car we have, sir, is a Ford Fiesta. It's a stick change. Can you drive a stick change?"

"Practically all we have in England," I said. "Really!" She looked surprised, "Wow."

We took care of the paperwork and she told us what bay the car was in. We carried our cases to it and loaded up. By American standards it was tiny, but just what we wanted. Easy to drive and easy to park. It had a fifteen-horse engine and could really shift, I found out later.

We found our way out of the car park. The only plan we had was that we were going south. You can't imagine my horror as I pulled onto the ramp to join the freeway. There were eight lanes of cars, all sweeping towards us. I managed to filter in and quickly realised that this was a baptism of fire.

I'd had far too much to drink to cope with this. Also, it was not an ideal time to be learning to drive on the wrong side of the road. I said to Ed in a really earnest voice,

"Now listen to me, Ed." "Yes, Des."

"Keep an eye out for a hotel on this side of the road," I told him seriously. "I don't care what it is as long as I can just turn straight into the car park."

We hadn't gone far when he spotted a large hotel. "Coming up on my side."

I indicated onto a ramp and swept around into a huge car park. I got as close to Reception as I could. I can't tell you how relieved I was to get off that freeway.

In Reception we went straight to the reservation desk. We didn't even know the name of the place, but it looked real posh. It turned out to be *The Los Angeles Hyatt.*

"Do you have a twin room?" I asked, "For one night only."

"Sure do, sir," he said, tapping away at something. "That'll be one hundred dollars cash in advance."

Christ Almighty, I thought. It was a fortune. We looked at each other. "We don't have that much cash," I told him truthfully.

"Credit card?" he asked.

It was then that I remembered the American Express card. I fished it out. I'd never used it before.

"Is that any good?" I asked.

"That solves the problem," he laughed and, with a wave of his hand, a porter appeared from out of the floor and disappeared with our cases. We got our key and chased after him to the lift, where he was waiting. We got to our room, and it was everything you would expect for a hundred dollars. We tipped the porter and checked the place out.

"That's knocked a hole in the budget," complained Ed.

"Yeah, but we won't have to pay until we get back," I said. "At least it's better than getting splattered on that freeway. It was a nightmare."

"That's for sure," said Ed philosophically. "After all, it's a once-in-a-lifetime experience and there's no point in penny-pinching. It's not as if we haven't got it. See what's in the mini-bar."

I was glad that he was of the same mind as me. There's no point in doing things by halves.

We might never be back here again. We checked the time. It was late afternoon here. Ed had the mini-bar open and was taking out a few miniatures and mixers. He liked Bacardi, so I had gin. He had coke, I had tonic. He mixed them and gave me my glass. We clinked them together.

"Let's have a great holiday, Des," he toasted. "Let's do this one for Phil." "To Phil," we toasted.

Lying on the bed, we watched TV for a while, then decided to sleep a few hours before dinner. We woke around eight, showered, changed and went down to the bar. I think the travelling was catching up on us, because after two New York Strip steaks and a bottle of wine, we were beat.

We charged everything to the room and had an early night. We wanted to be fresh in the morning. We had only a rough plan, with no hotel reservations booked. Our first destination was San Diego but, if we saw anywhere we fancied along the way, we would stop over.

• • • • • • • • • •

We negotiated our way out of the car park and through the traffic.

Miraculously, we were on a highway on the coast heading south to San Diego.

The first place we recognised by name was Redondo Beach. A roadside bar looked promising, so we parked and wandered inside. There were a couple of pool tables and a scattering of local girls and youths smoking, drinking and having fun.

Ed was first to the bar. This was our first proper American bar and we didn't know what to expect. You have to remember this is 1979 and the tourist invasion was yet to begin.

"Two large beers, please," Ed ordered.

"Coming up," said the casually-dressed bartender in shorts and T-shirt. "You guys from England?"

"Yes," said Ed, "Just arrived yesterday."

Two heavy glasses landed on the bar as I arrived.

"On the house," the bartender announced. "I don't think we've ever had any guys from England in here before."

We must have looked confused.

"Take a stool," he said, and then to the rest of the room, "Hey, you guys, say hello to my friends from England."

There was a loud murmur. Pool cues were dropped on tables and half the room converged on us.

"Hey, how you doin'?"

"You guys are from London! Wow." "How long are you here for?"

"My name's Joe." "My name's Jake."

It was handshakes and questions all round. We were made so welcome it was unreal.

Every time we finished a beer, someone bought us another. It was embarrassing. "Where you guys headin'?" someone asked.

We gave them a rough idea of our itinerary and said we were planning to hit San Diego that evening. He got some napkins from the bar and started writing.

"You gotta stay at Harbour Island, man. It's really cool," he told us. "There's an old paddle-ship moored there called the *Reuben E Lee*. It's a bar, restaurant and nightclub now. You just gotta go there."

There was a murmur of approval. Someone said,

"You should stay at the Travelodge. It's great there and not too expensive. Write that down," he told napkin man. "You'll love it."

"Johnny boy," someone shouted the bartender. "Call ahead and see if they've got rooms at the Travelodge, Harbour Island."

They were telling us all the best places to go when Johnny boy shouted, "Sure thing, they've got vacancies."

"There you go. You're all set."

We left our new American friends to the chorus, "Be sure to call and see us on the way back."

We said we would, but who knows?

• • • • • • • • • •

We found Harbour Island. And fell in love with it. The Travelodge was a large hotel on a beach overlooking the marina. The bar was decorated as a ship, with anything nautical displayed all over. It was the evening shift on the bar and the bartender was an older man with plenty of chat. His memories went back to the sixties and the first Limey invasion.

That was us. According to him. Or the Beatles, as it turned out. He said they were barmy days and he loved them. We couldn't argue with that. He was showing off a bit, mixing all sorts of cocktails for us to try. We hardly paid for any of them and they were really gorgeous. They were quite strong, too.

We were having a great evening, and getting slowly sloshed. No one had asked Ed for his ID. You need to be twenty-one in California to drink. The cocktail waitress was really pretty and very friendly. She hung around us chatting at the bar when it was quiet. She was fascinated by our accents and wanted to know all about us and England.

"It's my day off tomorrow," she said airily. "I'm going to the beach." "That will be nice," we said, slightly drunk.

"Yes," she replied, "I'm going with my girlfriends."

"I bet you're looking forward to that," we said. "Have a good time."

"I'm sure we will," she answered. "We'll be down on the beach by the lifeguard post."

We had another few drinks and opted for another early night. We were planning to go to Disneyland in the morning.

After breakfast we sat on the decking overlooking the marina and beach. It was a beautiful, sunny day.

"Where did that waitress say she was going to be?" asked Ed.

"Down on the beach somewhere, I think she said," I answered, looking about down below. "She said she was going to be with her girlfriends, didn't she?"

"I think we might have been a bit slow there, Des," said Ed, thoughtfully. "You reckon?"

I didn't want to admit it, but I think he was right. "Well, anyway, we're going to Disneyland, aren't we?"

"Yeah."

We didn't know how long we were going to stay at Harbour Island. We guessed at least a few days, so we unpacked and hung our weary clothes. It wasn't that far to Disney. We got onto

1.5 interstate and came off at Anaheim.

We were early and parked by the main entrance. It was recommended for "first-time guests" to have the guided tour. It began on Main Street and included general admission plus seven exciting attractions. It was ten dollars and lasted two and a half hours. We decided on that.

It turned out to be enough. It was roasting hot and I don't think we were fully over the jetlag.

The Jungle Cruise was great. Pirates of the Caribbean, Davy Crockett's Explorer and Bear Country were worth seeing. The Mark Twain Steamboat in Frontierland was worth the visit. But, to be honest, the place was so vast you could do with at least a couple of full days to do it all.

We passed on Fantasyland and Tomorrowland, and headed back to the car. There was no air con, so we opened the doors to let out the oven-like heat that had built up in the Fiesta. With the car slightly cooler, we drove with windows down back to Harbour Island. We had a few beers and went to our room for a sleep.

We were going to the *Reuben E Lee* that night.

• • • • • • • • • •

It was a straight drive from our hotel to the *Reuben E Lee*, about two miles away. We could see it in the distance, lit up from stem to stern with ropes of lightbulbs. Its huge wheel on the back was like a giant Catherine wheel.

It was fixed to a permanent mooring on its own jetty. Wooden steps led up to its entrance. A young man in a white shirt and bow tie greeted us at the entrance to the bar. It was very grand and last-century.

The bartender and floor waitresses were dressed as sailors. The girls in short, white skirts and blue tops. We sat at the bar and ordered two large beers. We sat taking in the ambience of the place. The beers came on two white napkins.

"Menu, sir?" the young bartender asked. "Please, mate," I answered.

"You guys from Great Britain?" the bartender said. "I love your accents."

"Yes, we've come from England for a few weeks travelling," said Ed, "And I must say how hospitable I've found all the American people. They seem to really like us."

"Well, we don't see many of you around here, sir," he laughed. "You're a little unusual." "Been called a lot of things, I suppose," said Ed, "But unusual! That's a new one."

We seemed to be getting off on the right foot. A cocktail waitress joined us.

"Hi, you guys," she said, all teeth, blonde hair and brown skin. "How you guys doin'?

You're not from around here, I guess."

"No," said Ed. "We're a bit unusual."

The bartender cracked up.

"You guys are funny, man. Have a drink on the house. Two more beers?"

It's a funny thing but, every bar we ever went in, we always got a drink on the house. I don't know if it was us or it's an American custom. It got like a challenge to us to see if we got one. We always did, wherever we went.

We were hungry and, after checking out the menu, we took a table in the restaurant. It was expensive. Our budget was taking a real hammering. We went for the Surf and Turf, steak and lobster, with a bottle of Californian Chablis.

I paid with the American Express card, leaving a ten percent tip.

"Are you visiting our nightclub?" asked the waitress. "It's admission-free to diners." "Certainly are," we both said together.

"I'll show you there, sir," the waitress said, "If you'd like to follow me."

We followed a pair of beautiful, long, brown legs to the bow of the ship. She took us through and showed us to a table. Like the rest of the ship it was wonderfully nautical, with a small stage right at the bow, with a four-piece American rock band playing great music.

"Cocktails, sir?" she asked.

"Two Mai Tais, please," I told her, "Crushed ice." "Of course, sir."

"What's a Mai Tai?" Ed asked.

"Not sure," I answered honestly. "I saw it on the menu and it sounded nice. You'll like it - Bacardi and dark rum and fruit juice."

Well, I was right about one thing. He liked it all right. So did I. Last call was two o'clock.

We had a couple for the road and left for the car.

It had been a fantastic night. I aimed the car home and made it safely to the Travelodge for a well-earned kip.

• • • • • • • • • •

After breakfast we made plans to explore locally. Nothing too stressful.

After taking a little advice from the receptionist at the hotel, we set off with a list of places to visit. There were two main places: Mission Bay, which we intended to visit first, and La Jolla (pronounced La Hoya).

Mission Bay offered first-class aquatic facilities. Paddle around tiny islands and lagoons in canoes, or sail in catamarans and full-size sloops. It is also the home of *SeaWorld,* but we gave that a miss.

The beaches there are remarkably unspoilt, stretching 27 miles up to elegant La Jolla. It's Windansea and Boomer beaches are marvellous for surfing. It's a very popular resort for artists and the La Jolla Museum of Contemporary Arts.

We found a bar on the sand and watched our tans start colouring. We made plans for the evening, which mirrored exactly the plans of last night, and ended up on the *Reuben E Lee*. We decided it was time to move on.

And this trip was a big trip. Las Vegas.

We checked out our map and did a route. It was back up to Los Angeles then out across the desert to Las Vegas.

• • • • • • • • • •

We checked out at 5am, without eating. We could eat on the way. We wanted to get as many miles as we could under our belt early. If we could beat the LA rush-hour into the city, getting out would be easier.

According to the map there was one more or less straight interstate to Las Vegas, I-15. Things went well. We were on I-15, a long, desert road, and starving. There was a fifties-style diner ahead and well off the road. We pulled in and ordered a huge breakfast, that came with as much coffee as you could drink.

Fully refuelled, we got back on the road and, although the limit was 55 miles per hour, I sometimes hit 90. There was hardly a car on the road.

We passed a sign for Death Valley, that was beginning to sum up the way we felt.

It must have been over 1000F in the desert. We had the windows down and were still dripping wet with sweat. There was a roadside bar and gas station ahead. We stopped for beer. Like two hobos in our wrinkled shorts and T-shirts. The bartender cheerfully said,

"Hi, you guys been to Vegas and lost all your money?" "No mate, we're on our way."

"Too bad," he said.

"Two large beers, please." "Comin' up."

"Thanks."

We drank them in two gulps, ordered two more and sat at the bar trading banter with the bartender. He had an uncle in London. Did we know him?

"Nice meeting you guys," he told us, slapping two beers on the bar. "Best of luck in Vegas. Have these on the house."

"Thanks," we both said gratefully.

We just couldn't get over how friendly these Americans were.

Thirty-one

If, as we claim, European dreamers go to America and American dreamers go to California, where do Californian dreamers go? Well, take Interstate Highway 15 from Los Angeles east across the Mojave Desert and the Devil's Playground, through lava beds and countless dry lakes and over the Nevada state line until you see a town rise out of the desert, blinking its joyously evil neon signs in twilight! Sodom and Gomorrah with air conditioning - Las Vegas.

Las Vegas is one of those myth-laden towns that everybody should see before they die. No amount of familiarity with all the films that have shown the meretricious marvels of this gambling paradise can quite prepare you for the shock of hot and cold reality. At Casino Centre downtown, the neon signs are so overpoweringly bright that midnight can look like noon in the Mojave.

Mosaics of monster dice, poker chips, roulette wheels and playing cards are embedded in the intersections of the main streets. Every lit-up window shows dozens of people frenetically tugging the levers of one-armed bandit slot machines. A sign says, "For every US coin you got, we got the slot". Other signs invite you to, "Cash your pay check here" or "Exchange any foreign currency." They'll take your money any way you want to give it to them.

Primed with this taste of things to come, we intended to head out to The Strip, between Sahara and Tropicana Avenues, four solid miles of hotels and casinos, each with a nightclub advertising America's top comedians and singers and a score of girlie shows.

You'll also see wedding chapels attached to little motels, where you can get married for 15 dollars, "Witnesses available two dollars and fifty cents extra," and spend your honeymoon next door. If it doesn't work, there are also offices along The Strip where you can get an instant divorce, at a slightly higher, unadvertised fee.

Even if you're not staying there, you can visit the most extravagant hotels on The Strip. These are the true monuments of Las Vegas, palaces of the Orient, Ancient Rome, Persia and many other fabled places that never existed except in the minds of the hotel and casino operators. The most outrageous is *Caesar's Palace*.

That was where we wanted to be and why we had booked a room at a small Travelodge motel not far from it on The Strip. Diana Ross was the star attraction, and it was her we wanted to see.

Looking like two desert rats, we dumped our luggage in the cheap but adequate room and found a bar next door. It was colourful and lively, with pool tables at one end and a bar that stretched the length of the room.

It was obviously not a tourist bar, and definitely not the place for two greenhorns from England. Tattoos, bandanas and studded leather seemed to be the dress order. There was a group of cowboys playing pool and plenty of attractive, scantily-dressed girls hanging at the bar. Neon beer-and-tobacco signs flashed and there were sports screens scattering the walls.

My hair, which I normally pride myself on, was greasy and curly and stuck flat to my head. Ed's black, curly hair was as big a mess as mine and I figured that, in our dirty, creased and sweat-stained T-shirts and shorts, we must have looked like a couple of vagrants. I excused myself to some biker types at the bar and leaned through them to order from a pretty bartender in cut-off jeans and T-shirt.

"Could we have two large beers, please?" I asked politely.

"Sure can, fella, comin' up," she said, with the usual perfect white-teeth smile. "You got ID?"

"I've got my driving licence," I said, a bit surprised.

"What about the baby?" she laughed. "You got his milk coupon book?" The bikers started howling with laughter. One said,

"She's just foolin' with you, man. Come on, Gina, give them boys a drink. They sure look like they could use it."

"I'm just kiddin' ya," laughed Gina, as she passed the two heavy glasses across. "You guys high rollers?"

Again this brought howls of laughter from the bikers.

"Not really," said our Ed, "We've come to see Diana Ross at *Caesar's Palace*."

"Well, good luck with that one," piped up another of the bikers. "That gal's a sold-out show for weeks."

"They could stand in line," said another. "What's that mean?" I asked.

"They get cancellations every show. There's a line waiting by the door hoping for a ticket.

Some get lucky. It's the only chance you got, so you just gotta take it." "Thanks," I said.

You know how it is with first appearances. Well we couldn't have been more wrong about this place. The girls, it seemed, were street girls, working girls making a living. We chatted to them the same as you do any girl you meet. They were neither proud nor ashamed of their job.

It was their chosen profession.

Las Vegas never stops. There's no clocks anywhere. The bikers and the cowboys and most of the other customers were all casino- and hotel-workers on time out. "Cooling down."

One cowboy, with a huge hat, challenged me to a game of pool. Egged on by half the bar and most of the hookers, I accepted the bet of five dollars. A pool player I am not. I had, however, played for the *Monica* pub team and, on a very good day, might win.

This must have been a very good day. You know how, sometimes, every shot you take is not the shot you meant but somehow goes down a hole? Well that's how the game went. I picked up the two 5-dollar notes and received a huge slap of congratulations across my back.

"You're sure one hell of a pool player," drawled the loser. "You want another game?" "Another time," I said, covered in glory. Quit while you're in front.

I rejoined Ed at the bar. He was getting on famously with a really pretty hooker. Ed had, had a fair few drinks and they were loving his "funny" accent and tired old jokes. They hadn't heard them as often as me. But we were sure having a great time.

"You can have one on the house if you like, hon," said the pretty girl.

"Thank you very much," said Ed, "But me and me bruv, Des, had better get some rest for tonight."

"Sure thing, hon," she laughed. "You come in again later?" Ed looked at me.

"You bet we are," I said. "See you later".

With a few cheerful goodbyes, we went next door to our motel. We had brought with us some really smart clothes for *Caesar's Palace*. Ed had some smart, expensive sports jacket and trousers. I had a really upmarket white suit with a dress shirt and green velvet bow tie. We got them ready for later and crashed.

• • • • • • • • • •

There were still a lot of the same customers in the bar, and more new ones too. Gina was still working and looked a bit confused as we ordered our drinks. A couple of the original bikers were there with a few fresh faces. One of the originals started to laugh.

"I never would have recognised you guys."

The piss-take was relentless but good-humoured and great fun. It came time for *Caesar's Palace* and they wished us luck.

We were overwhelmed by the gambling, hundreds of slot machines spilling out into the lobbies, plus roulette tables, baccarat, blackjack and bingo. The most enjoyable games to watch are the dice or craps games, where the players get excited, yell, whoop and groan as their fortunes go up and down.

Even if you don't gamble, the spectacle in the casino is fascinating and the floor shows in the nightclubs are first-rate. And it doesn't come much better than Diana Ross.

We found the entrance to the theatre. There was a huge foyer before a bank of doors.

There was a line of at least fifty hopefuls waiting for cancellations or touts with tickets to sell. It didn't look hopeful and I'm not much for standing in queues.

There were a couple of floor-walkers hovering about, tall blacks, who looked really fit. I casually strolled over to one.

"Excuse me, mate," I began, "But could you tell me where I can get two tickets for the show? Where's the box office?" Ed wandered up behind me.

"There ain't no box-office tickets, man," he told me. "This is a sold-out show. See all those folks over there in a line, they waitin', for cancellations but they ain't got no chance."

"Shit," I said. "We've come a long way."

"You shoulda bought tickets in advance, man. Specially you cummin' such a long ways." "Yeah," I agreed, "But we never thought. Oh well. Can't be helped."

Ed started moaning.

"We've come all the way from England to see her."

"It ain't my fault," I protested, "How should I know?"

"You guys from England?" said the floor-walker. "I thought you had funny accents. Look man, I ain't promising you nuttin', but you go stand over there a while. See what I can do."

"Thanks, mate," I said to him. "Yeah, thanks," added Ed.

We stood over by the wall away from the line, chatting amongst ourselves hopefully. After about 15 minutes, the floor-walker swaggered across to us.

"Got two tickets if you want them," he said, "25 bucks each. That OK?" "Sure thing," I told him, handing over the cash.

He walked us over to the doors and, after a few words with the doormen, we were through into a vast theatre with tables in tiered, circular rows from floor to roof. A hostess showed us to a table for two in a small booth and asked if we wanted red or white wine.

"Any other drinks, sir?"

"White wine and two large beers, please," I said. "Certainly, sir," she said, and left.

She returned shortly with two pints of lager and a bottle of local wine. The wine was complimentary with the table. I paid for the beer and gave a generous tip. I wanted her to keep an eye on us. She did.

"I say, Des, this is all right, ain't it?" said our Ed excitedly, drinking half his lager in one go.

"Yeah," I warned him, "Take it easy, it could be a long night."

A young, black, female singer came on with a great, lively band. She was really good. We were having a great time. We drank the wine between pints. The waitress knew the score and kept them coming.

Finally, Diana Ross came on and I have to say that what we saw of the act was class. She did all the Supremes stuff and left the stage to wander through the audience, chatting to people at their tables.

Unfortunately, when she got to our table, we were told, we were both asleep. Apparently, she had the whole house in stitches as she apologised, saying that she didn't normally send her audience to sleep.

"Too bad. You can't please them all."

The audience loved it.

• • • • • • • • • •

The next thing I clearly remember is sitting in our hookers' bar guzzling buffalo chicken wings, with grease and sauce all down my white suit. Ed was asleep on his arms next to me at the bar.

"What happened to you guys?" someone asked incredulously. "It's a long story."

We had flights booked for a tour of the Grand Canyon on the next day with Scenic Airlines. Non-refundable if we didn't show. I shook our Ed to life and half-dragged him to the room, where we collapsed on the beds fully-clothed. We had to drive to the airport for 9.00am the next day.

We felt like shit, but we made it.

Thirty-two

We had never flown in a light aircraft. We swore we never would again.

It was a six-seater. There were only three passengers, me, Ed and a beautiful young girl from somewhere in Europe, I would guess. The pilot, to us anyway, was an old geezer in an old, leather flying-hat. Biggles, I named him instantly. The girl sat next to him and me and Ed had the rear to ourselves. The take-off could only be described as terrifying.

Biggles banged off down the runway and we wobbled from side to side up into the air.

Biggles gave a running commentary that I think he may have given a thousand times before. I was on the edge of my seat, threatening to be violently ill with every swoop and turn. Ed didn't seem too bad. I had a sick bag held at the ready under my chin. I've never felt seasick in my life, but this was horrendous. I was fighting against last night's booze and chicken wings all the way.

Biggles seemed to have taken a shine to the girl and was holding her bare, brown thigh a lot tighter than that joystick. We swooped over the Hoover Dam, a spectacular engineering experience, its huge pylons creating a surprisingly attractive sculptural effect across the landscape, I was told.

Bothered?

Banking away from that, we flew like the Dambusters in and out of the Grand Canyon, miraculously skimming the sides of mountains while Biggles held on to a bare, brown thigh as if heroically preventing its owner from falling out of the plane.

A wonder of the world. Obviously showing off, the only things he didn't do was a loop the loop or fly upside down.

One of the Seven Wonders of the World, I vowed to myself to see the other six in front of the telly, with a four-pack and David Attenborough.

I managed somehow not to throw up but, if you thought it couldn't get any worse, there was the landing. Staring over Biggles and his bird, we watched through the windscreen as we made our approach. The plane smacked down in the middle of nowhere.

There were a few shops and buildings, and a car park with a few coaches. We looked at each other, me and Ed, and both agreed.

We needed a drink.

Biggles cheerfully informed us that there were no bars in the Grand Canyon. We disembarked and asked when we would be going back, thinking it couldn't take long to look around a couple of shops. Biggles said he would be back in three hours.

"What do you mean?" we asked. "You're booked for the Deluxe tour."

"What's that?"

"You get to sightsee on a coach for a couple of hours," he informed us.

"Well if it's all the same to you," we told him, "We'd just rather go back with you." "Sorry, no can do," he said, "The seats are fully booked for the trip back. You're booked on the next return flight." And so it was.

Two friggin' hours driving around mountains, rocks and valleys. I can tell you now, when you've seen one mountain you've seen them all.

We had to draw the line somewhere.

The Deluxe tour included a half-hour stint on a horse. Well, trust me, the plane ride was bad enough, but there was no way we were bouncing around this Wonder of the World on a bleedin' horse. We sat on a bench and waited for the more adventurous explorers to make their triumphant return.

Biggles came back for us as he had promised and, after another terrifying flight through the wilderness and canyons, we arrived back at Las Vegas. He gave us a certificate to say we had had the memorable experience of being one of the few who had flown around the Grand Canyon. Lest we ever forget.

It was then, and only then, that we realised the sacrifice made by the early pioneers and settlers of this great country.

We shared our experience with our new-found friends, the bikers, pimps and prostitutes of Las Vegas. They were all thoroughly amused and, as veterans of the West, were hard-pressed to see why we had bothered in the first place.

Ed's girlfriend was very sympathetic, telling him,

"You can still have one on the house, hon, if you want it."

He went missing at some point in the festivities for about half an hour.

"I was just getting some air, Des," he told me, that impish, innocent smile of his all over his face. That was Ed through and through. He loved the mysterious, and would never say if he had or he hadn't.

I have my suspicions.

• • • • • • • • • •

The next thing on our to-do list was San Francisco. We thought with an early start we could do it in a day. We had flights booked to Honolulu on 17th Sept, return 22nd, giving us five days there. Checking the date, we had three days to explore San Francisco if we got there in one day.

We were used to getting up early all our lives and liked it that way. The Old Man had always told us,

"An hour in the morning is worth two in the afternoon." We lived by that rule.

It was dark when we started out, and cold. For the first time, we had the heater on in the car. It's true what they say about the desert. The only thing worse than the heat of the day is the cold of the night.

The only car on the freeway meant that I could floor it most of the way. The car was a dream to drive, and ninety-five miles an hour was no problem for it. You can cover a lot of ground. We were burning up the miles.

And the early start had an exciting and unexpected bonus. Sunrise in the desert. It rose slowly into the sky, a ball of orange flame, huge and awesome. Wonderfully spectacular.

I can't remember what time we got to LA, but we found ourselves trapped in eight lanes of traffic heading into the city. It took a long time before we found our way onto the Pacific Coast Highway. We were hungry at this point and started looking out for somewhere to eat. The original plan had come to fruition and we were past LA. Now we had the rest of the day to take it easy to San Francisco.

The Pacific Coast Highway is probably the most famous road in California, so we wanted to take it all in. We saw a sign for Oxnard. It looked small on the map, so we stopped there and found a roadside diner serving breakfast. We had a full breakfast. Steak, over-easy eggs, hash browns and all the trimmings with copious amounts of coffee.

That would do us for the day. We started off again for 'Frisco. The highway lived up to all the hype. The ocean to our left and spectacular views to our right. We passed Santa Barbara and Lompoc; they looked too large. We stopped at Santa Maria, Morrow Bay, San Simeon and Carmel. It was mid-afternoon. We'd had enough. We weren't going to get to San Francisco.

We wanted beer.

A place not on our list popped up on one of their big, green, overhead road-signs. San Jose.

We'd heard the song *Do You Know the Way to San Jose*?

We took a right and pulled into the car park of the San Jose *Hyatt*.

After the cheap motel in Las Vegas we wanted some luxury. Pioneers and settlers, we had decided, we were definitely not. As is often the case, the unexpected turns out to be the best. We just about made our flight three days later.

Thank God for the American Express.

• • • • • • • • • •

The rooms were to die for. Massive bathroom with thick, sumptuous towels and robes, TV, fridge, mini-bar and balcony overlooking a fabulous pool area with beach-shack-style bar and comfortable sunbeds in spacious grounds. We showered and changed into our swimwear and hit the bar.

Problem. The bartender wanted Ed's ID. He hadn't got it, of course. However, I cleared things up when I ordered water for Ed and a pitcher of beer for myself. We found a secluded bar area amongst palm trees, where Eddy watered the plants and helped himself to some beer from the pitcher.

"I hope we're not going to have this problem in the hotel," Eddy said, "Or it ain't going to be much fun for me."

"Don't worry," I reassured him, "There's always a way."

We drank away the remainder of the afternoon with glasses of water and pitchers of beer. I'm sure they must have wondered where I was putting it all. But, truthfully, I think they knew and turned a blind eye. I'm sure they were within the law as long as they didn't serve Ed.

When we'd had enough sun, we decided to rest in the room. I ordered a pitcher to take up with us and signed for the bill. The drinks were so expensive, we decided to worry about it when we got back.

We rested up, watching some American football. Ed used to watch it at home and knew most of the rules. He educated me and I was getting into it, too. It was time for dinner, so we dressed up in some of the smart new clothes we had bought especially for the holiday. Nice, casual jackets and expensive trousers and shoes. We had gone into Birmingham one afternoon to my favourite shop at the time, MAX. It was a small shop in New Street that specialised in nice, Italian clothes. We had spent a fortune, but it was worth it. Needless to say, I negotiated a very reasonable 25% discount. They were still earning good money out of us.

We were escorted to our table by a smart, friendly waitress. Extremely professional, she guided us through the menu and the specials of the day. We ordered clam chowder soup and two New York Strips with all the trimmings. A bottle of Californian Chablis and we were "all set".

It was quiet in the hotel, we were told; apparently it usually was at weekends. It was full of business people in the week. The soup and steaks were wonderful. The waitress asked if we wanted sweet and coffee. We politely declined and sat comfortably chatting to each other. A band and female singer were setting up in the bar.

"I think we should sort a table out in the bar," I said to Eddy. "We need one out of the way where you don't stand out too much."

"Yes, Des," Ed quickly agreed. "I can't go all night without a drink."

"Don't worry, I'll sort it," I reassured him. "Tell you what; you stay here and I'll go and suss it out."

I left him there and made my way into the bar. It was pretty quiet. I looked around the room and there were a few tables against the far wall that overlooked the band. The lighting was quite subdued and I found the perfect spot.

The bartender was polishing glasses behind the bar. I sat on a bar stool. "Hello, mate," I greeted, "How are you today?"

"I'm good, sir," he answered, in a bright, friendly, voice. "And how has your day been today, sir?"

"Marvellous," I said. "A little tired, though; we drove from Las Vegas today; long trip." "Gee, man, that's one hell of a drive."

"Yes; we started out while it was still dark," I told him.

"You must have hit the floor heavy with the gas," he said, with admiration. "You got a Ferrari or something?"

"A Ford Fiesta."

"You're kiddin' me," he said, placing my beer on a napkin. "Hey, Yvonne, you hear this?

This guy drove from Vegas in a Ford Fiesta. Ain't that a stick change?" "Yes," I said.

Yvonne was sitting on the end stool with a cocktail. She was immaculately dressed in a grey suit and cream blouse. In high heels, she slid elegantly off her stool and gracefully joined us. The badge pinned to her ample breast identified her - Yvonne, Assistant Manager. She was five feet ten, slim, perfectly straight, white teeth set in a beautiful, brown, soft face, curtained with shoulder-length blonde hair and large, sparkling, ocean-blue lagoon eyes.

She would have made a welcome guest in anybody's fantasies.

"You must be one of the British guys they're all talking about," she grinned.

"Well, we're British all right," I owned up. "Whether we're the ones they're all talking about, I couldn't say."

"You have to be," she said, sitting next to me. "I can't remember ever having any British guests before. Are you on vacation?"

"Yes," I told her, "With my brother. We're touring around for a few weeks." "How wonderful. Where is he?" she asked.

"Finishing his dinner while I find us a table for the evening," I said. "I've found one over there and I'm getting a drink sorted out."

"Great," she said. "I'm sure Tony can fix you up. Hey, Tony?" "No problem; what's your poison?" Tony asked.

"A pitcher of beer and a jug of water for my brother, please."

"Sure thing, comin' up. I'll get one of the girls to serve it to your table."

"Thanks," I said, holding out my hand. "I'm Des, my brother's Eddy." I shook Tony's hand and then Yvonne's.

She had a soft, warm hand. I took my time releasing it. She noticed. "I'm Yvonne," she smiled.

"I already got that," I told her. "I'd better get my brother. I'd like to talk to you a little more."

"I'd like that too, Des. Would I be interrupting you if I joined you a little later for a drink?" she said.

"Probably, but I'll make you some room." We both laughed. I think I had a date.

"Where have you been?" complained Ed. "I've been here ages."

"Getting in with the management," I explained, and told him what I'd been doing.

When we got to our table the beer was already there, and water for Ed. Ignoring the water, we poured some beer. The band was warming up and the singer was really good. We were having a great evening, getting drunk and planning the rest of the trip. We were very excited about Honolulu; it was somewhere I had dreamed of going all my life. Ed too.

It was about 11.00 when Yvonne joined us. She had with her an exotic cocktail. We both stood up and I helped her with her seat. She seemed impressed. She complimented us on our lovely European clothes and said how lovely it was to have us as guests at the hotel.

We all asked the usual inquisitive questions of each other and I told her I was a diver and did quite well for myself salvaging old shipwrecks. Fascinated, I slightly exaggerated how there was untold wealth at the bottom of the sea for those prepared to take on the risks of salvaging it. I told her of my close shave with a tiger shark a few years earlier while trying to locate the wreck of a Cubana Airways plane that had crashed after take-off in Barbados.

Yes. She remembered it. It had made big news in America. Was it the work of terrorists? I told her truthfully that I didn't think we would ever find out. We couldn't find the plane and concluded that it would remain lost for ever in the extremely deep water it was suspected of being in.

Equally fascinated by this story he had never heard before, our Eddy listened in awe. He couldn't believe I'd never said anything about it before.

"I'm a bit tired, Des," he announced. "Do you mind if I go to the room? I could do with a good sleep."

"No, Ed, you carry on," I said. "I'll see you in a bit."

"He thinks a lot of you, doesn't he?" Yvonne said softly. "We look out for each other."

I told her the story of our Phil.

"Bloody hell, Des," she said, reaching across the table to hold my hand. "That was a tough hand to be dealt."

"Yeah. Well let's not talk about that."

Last call was shouted. I attracted the waitress. "Two more, please."

The band packed up and left. The lights went down. Tony said goodnight. We sat in semi-darkness talking and laughing like we'd been friends for ever.

"Do you like wine, Des?" Yvonne asked. "Who doesn't?" I asked back.

"Are you in a hurry to go?" "Not while you're around." "Wait here. I'll be back."

When she came back she was carrying an ice bucket and two large wine glasses. "Compliments of the house."

We sat in the empty room until well gone 2am. We kissed goodnight and she said it was her day off tomorrow and what were we doing?

"Resting by the pool. If you've got no plans, why don't you join us?"

"I'd like that," she said, "I'd like that very much. But won't Eddy mind?"

"You must be joking," I laughed, "He likes you and I think he needs a change of company.

You'll have to be prepared for his corny chat-up lines, though."

We stood up and walked to the door leading to Reception. We stopped just short and kissed quite passionately. I took a chance and pulled her into me. She didn't object. We broke apart. We looked once more into each other's eyes and did it again. A little breathlessly, she told me,

"See you tomorrow, Des." "You too, Yvonne."

It didn't seem right leaving her. We were in a half-empty hotel, for God's sake. Maybe tomorrow.

• • • • • • • • • •

We've all heard about the great American dream. I suppose everyone has a different one. At that moment, sitting in the sun with a glass of freezing beer, by the pool, I knew what mine was and hoped it was within my grasp.

Yvonne. She was walking gracefully as a catwalk model across the almost-empty pool area towards me. In a white, not-too-skimpy bikini, her long, brown legs swallowed up the yards between us until she was standing by my table. In fashionable sunglasses and with hair tied behind her in a plaited ponytail, she looked as if she should have been on the cover of some beauty magazine.

"Hi Des. Eddy," she greeted us cheerfully, "Mind if I join you?" "If you must, I suppose," I said grudgingly.

She knew I was joking and sat next to me.

"Too bad if you don't like it," she laughed. "Anyway, I wasn't talking to you. I was talking to Eddy."

That was the way it went for the morning. Beer and glasses of wine kept appearing apparently from nowhere and, around two o'clock, a platter of buffet-type food landed.

"Left over from lunch, I guess," she explained, with a mischievous grin. A bottle of white wine arrived in a bucket, with three glasses. It was only about three-quarters full.

"Believe it or not," Yvonne told us, "People order a bottle of wine, drink one glass out of it and leave the rest. We never let it go to waste."

I was in full agreement.

"Is everything all right with your room, Des?" Yvonne asked, with an almost undetectable wink.

"Well, now that you mention it," I began, "There are one or two things I'm not too happy about, but you're off duty."

"You're never off duty in this job, Des," she said. "Show me and I'll get it sorted out." She stood up. I got up too, and turned to Ed.

"I'm just going to sort out those problems with the room." He raised his eyebrows, rolled his eyes and said,

"Yes, Des. And don't forget to mention room service."

Stepping into the room, I was just about to close the door when Yvonne stopped me.

Slightly tiddly, she raised her finger to her lips and slipped the DO NOT DISTURB sign onto the handle outside.

Only then did I realise I had been tricked. Happy days.

• • • • • • • • • •

We had one more day to go. After sorting out room service with Yvonne, me and Eddy had dinner and drinks. Yvonne joined us at last call and Eddy excused himself again. A couple of hours later I got to our room.

There was no sign of Eddy. I was slightly worried, but could not think of any harm he could be in. However, I couldn't sleep until he was back. I switched on the TV and had a beer from the mini-bar. About an hour later there was a knock on the door. I opened it, and hotel security was supporting a very drunk Eddy.

"Do you know this man, sir?" the man in black asked. "Yes, he's my brother," I answered, "Where has he been?"

"In another guest's room, sir," he said, with a hint of a smile, "Trying to empty the mini-bar, I think."

"Is everything all right?" I asked.

"No complaint has been made, sir," he told me, "But when it was time for him to go, she couldn't get rid of him. He couldn't find his way, so she called us and asked us to escort him back to his room. Here he is, sir, he's all yours. Have a good night now."

Ed crashed in and onto his bed.

I'd get the story tomorrow.

• • • • • • • • • •

He told me all about it by the pool. He was going back to his room when he met an attractive woman in her thirties who had seen him by the pool and noticed he seemed alone. She was on some weekend break and had had no company all day and was feeling slightly pissed-off. Did he fancy a couple of drinks in her room? He did, of course, and, when she wanted him to leave, he was a bit the worse for wear and could only remember his room number. Security rescued him.

Our last day fled. Yvonne was working, of course, but managed to see me for an hour on her break. We sorted out the recurring problem with the room and, after dinner and drinks that night, we said goodbye.

We were leaving early in the morning to get our flight to Honolulu. There were no tears or sadness, just joy and gladness that our ships had passed in the night. We both agreed it had been great fun and knew that we would never see each other again.

We didn't even bother to exchange details. What for?

Thirty-three

We'd parked in the appropriate car park and were in line at the desk for Honolulu. Our turn came quickly. It was a domestic flight. All seemed to be going well. Then the girl scrutinising our passports and tickets glanced across to her colleague. Silently, they seemed to agree to something.

I didn't like it. I smelled a problem.

I looked at her. She looked at me. She looked again at our documents. "You guys have travelled from Great Britain," she told me.

"Yes," I said. "We're trying to see as much of your wonderful country as we can while we're here. And I would like to say how wonderful everyone has been to us."

"Thank you, sir," she said, "I'm very glad to hear that. However, we have a problem." "I'm sorry," I said. "What's that?"

I feared the worst.

"Your flight is overbooked." "Shit," I said, without thinking.

"I'm afraid, sir," she began, "I am going to have to upgrade you to first class.

Is that all right?"

I was a little shocked. The girl was trying to contain herself but, after a glance across to her colleague, they both started laughing.

"I'm sorry, sir, but I have to press you for an answer. Is that all right?"

"Oh, go on then," I laughed, "Thank you very much. That's very kind of you." "You're most welcome, sir," she said, handing me my boarding cards.

"Have a nice flight. I'm sure you will."

She took our luggage and we thanked her again. "Bloody hell," gasped our Ed, "What happened there?"

"Just another star-struck woman wilting under my irresistible charms," I told him truthfully. "You've either got it or you haven't."

"Yes, Des." He didn't bother saying it. He'd heard it all before.

"What?" I protested.

• • • • • • • • • •

I'd never flown first class before and, as it was to turn out, I never did again. As wonderful as it was, I'd rather have the couple of grand extra it cost in my pocket, to spend when I get there.

The first thing we had was a glass of champagne. It was all right, but we'd rather have beer. "No problems, sir. Two beers coming up."

With the two beers came the cocktail menu.

"Perhaps you'd like to try one of our cocktails, sir?" she said.

Worried that we may have offended her with the champagne, we had a couple of Pina Coladas. With those came the food menu.

There were only about four choices, but any one of them would have been at home in the *Hyatt*. I can't remember what I had, but it came with two small bottles of wine.

The flight went by in a flash and soon we were walking down the steps onto the tarmac.

Just like in the films, beautiful Hawaiian girls with grass skirts and black, flowing hair placed garlands of flowers over our heads.

We were here for five nights and did not intend driving. This was relaxation time. We took a cab and told the driver, "Holiday Inn, Waikiki Beach." We passed Pearl Harbor and made a note to visit it. We never did.

The hotel was all you could wish for. We paid extra for an ocean view. It was well worth it.

We were something like twenty floors up, with a big balcony overlooking the ocean. It was late afternoon and the breakers rolled in. The surfers stood on their boards, criss-crossing and skimming the surf. I couldn't wait to show them how it was done properly.

The first night we ate in the hotel. The food was good and reasonably priced. We didn't feel like travelling far, so decided to stay in the bar. It was really quiet. We took stools at the bar and ordered two large beers. The bartender was young, about twenty-one or two, I guessed.

"You guys on vacation?" he asked.

"Yes," said Ed. "We're from England. We're doing some travelling around California. As we were so close, we thought we'd have to see Hawaii. In fact, if you don't mind, we'd like to pick your brains on the best places to visit."

"Sure thing," he said. "No problem. Do you guys surf?"

"No," we told him honestly, "But we wouldn't mind trying it."

He looked very sceptical.

"This isn't the place to learn, man. This place is, like, for the serious dudes. Have you seen those breakers? It's dangerous out there. Not for beginners."

"We have watched the surfers this afternoon from the balcony," I said, "And I see what you mean. We thought about going to the out islands."

"What for, man?" he laughed. "They're no different to this one other than there's absolutely nothing to do when you get there. They're for old people to sit and gaze at. You young guys want some action. Am I right?"

"Well, yes," said our Ed, "So where do we get it?"

We ordered two more beers and he began to give us the rundown on the best local bars.

There were a few good nightclubs close by. In fact, for the short time we would be here, he said we were in the best area. Taxis are really expensive. So why bother moving when there are plenty of places local? We agreed with him.

"The best nightclub anywhere is right next door," he told us. "*The Point After*. The only trouble is you can't get in unless you're Asian or known to them. They don't have many Americans in there, nearly all Orientals. It's like their club."

"Well, that's a shame," our Ed said. "It sounds really nice."

"It is. Best club on the island. Thing is, you guys are different. You're not American, so it might be different for you. Who knows? I've never met anyone from England before and I bet they haven't either. I love your accent. They might, too. Try it out."

We ordered more beer.

"Can we buy you one?" I asked him.

"Sure," he answered, "Thanks. I'll have a Micky."

He knocked the top off a bottle and took a draw. He poured us two frothy pints. They seemed to like a big head on it in America, we noticed.

"Can I ask you a question? he asked. "Of course."

"Is it true they drink warm beer in England?" Ed and I exchanged grins.

"Yes, but they are starting to drink lager," I told him. "A few people are drinking it, but they don't like the taste and have lime in it. I don't think it will ever catch on, myself."

A few other customers drifted in. Most sat at the tables and were served by the cocktail waitress. The bartender fixed the drinks and she delivered them to the table.

"Have you guys tried our Polynesian speciality?" he asked, "Pina Colada." "Yes, we've had one somewhere."

"Have you had one here?"

"No."

"Would you like to try one?" he asked. "I think you'll find it different." "Why not?"

He disappeared for a couple of minutes and when he arrived back he was carrying two giant pineapples on stands. They were hollowed out and filled with booze and decorated with straws and small umbrellas. They looked amazing and tasted fantastic.

But what a waste of a perfectly good pineapple.

Before we left for bed, he told us one more thing that he said was very important. The street we were on was Kalakava Avenue. The one side of it ran all the way along the beach. The beach side had a small wall. He told us never to walk along that side of the street. Stay on the side of the shops and bars. At night, it was not unknown for carless tourists to be dragged over it onto the beach and robbed. There were apparently gangs of Samoans who specialised in that sort of thing.

We thanked him for the information and charged the drinks to the room. We left a ten-dollar tip. He seemed pleased with that.

Back in the room, we watched some American football with drinks from the mini-bar and discussed the holiday so far. The American Express was taking a big hit, but we weren't changing so many travellers' cheques. We decided to change more cheques, as it was expensive to cash them back.

Reading the hotel information book, the poolside snack bar looked a good bet for tomorrow. So that was the plan.

After breakfast, towels over our arms, we found a couple of loungers by the pool and spent the day sunbathing and calling the poolside waitress for beer and snacks. We decided what we would do that evening. Have a sleep, eat and try *The Point After* nightclub.

• • • • • • • • • •

The club was situated in the large mall on the other side of the road in the *Hawaiian Regent Hotel*. Standing outside the entrance were two small, oriental men about twenty-five, in smart suits. We were dressed casually.

"Good evening," I greeted them cheerfully. "We'd like to have a drink. Is that all right? Or do you have to be members?"

Ed said, "We've been told it's a very nice place. We're on holiday and would like a drink." "Yes sir. Thank you. It is a very nice place," one of the doormen said. "It is ten dollars each admission, with the first drink free. Is that all right?"

"That's great," I said.

"Fine," said Ed, reaching into his pocket. He handed over twenty dollars and they escorted us up some deep-carpeted stairs to another entrance.

The main area of the club was large and spacious. There were three bars and a large dance-floor. It was decorated out as a sports bar. The usual flashing lighting and large sitting areas.

There was a restaurant area at the far end. It was quiet at that time of the night and was all oriental men and women.

The staff were all oriental, too. The waitresses were small and slim and walked the floor with their trays balanced on the palms of their hands.

Our escort from the door asked where we would like to sit. We told him we preferred to sit at the bar if that was OK. He said it was fine and showed us to one of the smaller bars.

"These gentlemen are from England," he said to the bartender. "Give them their complimentary drink, please. Enjoy."

"Thank you," we both said, settling on stools. "Two large beers, please."

We sat with our beers and surveyed the club. It was nice. A DJ was playing music and talking the talk. We got talking to the bartender and found he was very friendly and inquisitive. Over the evening, many of the staff would come and talk to us. Everybody seemed to be somebody's brother or cousin.

The manager came and introduced himself. His name was Ricci. He bought us a drink. They treated us like honoured guests. We had a really great night and, when we left, Ricci saw us to the door, shook our hands warmly and wished us goodnight.

We both agreed what a great night it had been and decided that's where we would go tomorrow.

The next day was spent by the pool again. Eddy had a habit of being too familiar with the poolside waitresses. He was a bit of a nuisance when he'd had a drink and tended to be a little excitable. I tried telling him to calm down a bit, as I could see that he was beginning to annoy one girl in particular. It came to a head when she refused to serve him any more. It was getting on anyway, so I apologised on his behalf and guided him back to the room. He flopped on the bed and slept.

After dinner and a drink at the bar we made our way to *The Point After*. It was my turn to pay the entrance and I got my money out.

"No, sir," said the doorman. "No entrance charge tonight. Ricci says you are his guests this evening."

Very flattered, we thanked him and made our own way into the club and up to the bar. The bartender saw us coming and two beers were on the bar for us. I handed over the twenty bill I still had in my hand.

"No, sir. First drink is complimentary."

"But we didn't pay to come in," I protested.

"Ricci says no matter," he said, waving away my money. "You are his friends from

England."

"Tell him thank you very much," I said, genuinely touched. "It's very kind of him." He bowed very slightly and went about his work.

"Blimey, Des," said Ed, "We seem to be well in here."

"Yes," I said. "It's nice of them and all that, but I really prefer to pay. It makes me feel a bit uncomfortable."

"It's only a bleedin' drink, Des," Eddy told me. "Don't start getting paranoid on me."

He was right, of course, but it's just the way I've always been. Another two beers arrived for us and, as I went to pay, the bartender said, pointing,

"My cousin bought you that."

A friendly face smiled along the bar and the man raised his glass to salute us. We raised ours too and drank. Throughout the night the same thing kept happening. Somebody always wanted to buy us a drink.

When it came time for us to go, a crowd of well-wishers saw us off. "What was all that about?" I asked Eddy.

"I think they must like us, Des," was all he could think of to say.

• • • • • • • • • •

I wish I could say the same thing about the Holiday Inn.

Next morning, there was a letter under the door of our room requesting us to go to the Reception desk. After breakfast we wandered over there.

They informed us politely that, as our room had been previously booked by someone else, we would have to vacate it. However, they informed us, they had negotiated the same rate for us at the *Hawaiian Regent* across the street. Did we need a cab?

I told them no. We could walk. We were politely being thrown out. We thanked them for their hospitality and went off to pack. Obviously, the poolside waitress had complained and we'd got the push. It was a bit of a slog across the street using the underpass and, in the heat, we were sweating a bit when we checked in to our new accommodation.

It was hassle I could have done without but, when we saw our room, we were well impressed. It was far grander than the Holiday Inn and also had an ocean view. It was also a lot more convenient for *The Point After*.

We found the pool. And regrouped. We decided to have a good drink in the afternoon, get a good sleep and get to the club a lot later so that we could stay longer and see the place when it had really livened up.

That's what we did. We partied until closing. It was two in the morning. The DJ was calling for everyone to leave. The place started to empty. Taking a hint, I said to Ed,

"Time to go."

We got down from our stools and began saying our goodnights. Ricci came rushing over.

"No, no, no. You don't leave," he told us. "You are my friends. Stay and have a drink with the staff. It is for us the best time of the evening."

The club emptied and, one by one, the staff gathered around the bar, drinking and talking. Somebody always seemed to be fussing around us, asking if we were all right. Large plates of snacks were laid along the bar. We were told to help ourselves. We did.

By 3.00 am we were three sheets to the wind and, fending off all protests, said we had to go.

Ricci intervened. He told us that at this time of the night it was not safe to walk through the mall to the hotel. He made a phone call and escorted us to the exit. Outside the door, we were met by an armed security man in a light-brown uniform. He was black, and Ricci gave him instructions to see us safely to our room. He followed us, a yard behind, to the hotel. We thanked him, but he insisted he saw us safely to our room. That's what Ricci had told him to do and that was that.

We had two days left and realised that we hadn't seen any of the island. So, the next day after breakfast, we went exploring. We wandered a couple of miles along the promenade, stopping here and there at a beach bar for a beer and to watch the surfers.

There was a massive shopping mall and streets of shops to wander around. We spent the whole day exploring, then had some food and a sleep to freshen up for another night in the club.

It was our last-but-one night and the party was going well. Ricci was even more attentive than usual.

"I've never seen you dance," he said. "This is a nightclub - you should be dancing."

"We don't really dance," I said, honestly, "And we don't have anyone to dance with anyway."

He shrugged as if he understood, and I thought no more of it. That was until he returned with two pretty girls. One was typical blonde LA.

The other was mixed-race Caribbean-American, light-brown chocolate.

Announcing to the world, he said to the girls,

"These are my friends from England. I want you to be nice to them."

Ricci bought them a drink and we got on really great. They were really friendly and great fun. The drinks kept coming and the girls kept drinking them. It was time to go and Ricci got the girls another drink. We asked if they fancied a drink in the hotel. Ricci enthusiastically encouraged them, saying all the time what great friends of his we were and to make sure to be nice to us.

The security guard was called to see us home. I said the door of the hotel would be far enough. It was agreed. We found our rooms, had a few drinks from the mini-bar and watched some American football. The girls left before breakfast and found their way home.

· · · · · · · · · ·

The last day was the same. Although we agreed we'd had a fantastic time in Honolulu, we also had to admit that we hadn't seen much more than Waikiki. Five days is not long. We could have done more but got carried away on the partying. Would we have done it differently?

Probably not.

The last night in the club, we were made to feel even more special than usual. Ricci asked if the girls had been OK; I said yes, they were fine. When they had had their fill, they left and said what a great night it had been.

We had afters with the staff and Rick presented us both with two boxes of presents. There were glasses from the club, souvenirs of all sorts, miniature drinks and a coconut each. He gave us two of the club's business cards and, on the back, he had written that we were both honoured life members.

We felt a little sad as we left for the last time with our armed security guard. He shook our hand vigorously at the hotel entrance and wished us a good life. We wished him the same.

· · · · · · · · · ·

We landed back in San Francisco in the early hours of the morning. We had five days left of our holiday. We'd planned to spend a couple in 'Frisco.

I don't know what happened, but we got totally lost. I guessed that we were somewhere in the downtown area. Wherever we were, there were vagrants standing around burning oil-drums drinking and shouting. We drove around, hoping to find a friendly sign that might get us away from there.

We were frightened. It seemed as if we were attracting attention that we didn't want. I think we both saw it together.

On the overhead gantry, the green road sign said San Jose.

It only took one look at each other and we were following the arrow away from this shit-hole. It was daylight when we saw the high-storey building against the skyline. Knowing where we were, we saw a Denny's Diner with a neon sign advertising breakfast.

We ate our fill and spent the next few hours wandering around San Jose, checking out the shops and bars. Finally, early afternoon, we checked into the *Hyatt*.

For one night only.

There were still a few more places to see and, after today, only four days to do it.

• • • • • • • • • •

We were sitting at the pool bar when Yvonne found us. "Des, Eddy!" she exclaimed. "I heard you were back." "Good news travels fast," we laughed.

"I thought that was bad news," Yvonne countered.

"Well, perhaps it is bad news," I said. "How do I know?" She leaned in to me and gave me a huge, soggy kiss. "Well, what do you think now?" she asked.

"I'll give you the benefit of the doubt," I surrendered. "If you want to double check, there's a gun in my pocket."

"Well, hold your fire until I'm ready," she teased me. "When do you finish?" I asked.

"I'm on earlies today. I finish at six."

"Ed. Do you mind if Yvonne joins us for dinner?" I said and meant it. I would know if he did.

"Des, I don't mind at all," he reassured me. "You know how much I like Yvonne. I've always said what lovely come-to-bed eyes she's got. It's just a shame they're come-to-bed Desi eyes and not come-to-bed Eddy eyes."

We all laughed as I nearly shoved him off his stool.

"Not that old one again," I sighed, with boredom. "I think you need a new scriptwriter!"

We never stopped laughing all night. All through dinner and drinks afterwards. It came to that awkward time in the evening when we had to think of bed.

Yvonne turned to Eddy and gave him a big hug and a kiss on the cheek. Retrieving a key from her purse she said,

"I hope you understand, Ed, but it's our last night together." "Not another one."

Dangling the key before me, she said, "Perks of the job."

• • • • • • • • • •

I phoned Eddy early in the morning.

Yvonne was gone. Earlies. I got back to the room and packed, loaded up the car and caught first breakfast.

We took the sign for Santa Cruz and picked up the Pacific Coast Highway. We had to decide where we were going. We had four nights left and still a few to-dos left. One was Tijuana, over the border in Mexico. We just had to do Mexico.

Tijuana was only a few miles from San Diego. We decided that the obvious place to stay was Harbor Island. It was a long drive but, if we kept the beer stops to a minimum, we could easily be there by evening.

We stopped for beer at San Simeon, Santa Maria, Santa Monica and San Clemente.

Finally, the Travelodge, Harbor Island.

Luckily, they had a room and we spent the evening in the Captains Bar telling the old bartender all about our adventures. He was an avid listener. We told him we were going for breakfast in Mexico the next day. He warned us to be careful.

"They're all crazy down there."

We took in his warnings and left early in the morning for the border. The checkpoints stretched fully across the road. Booths, with American guards in each one. There were about eight. Four in, four out.

They waved us through after checking our passports. What we were about to see was possibly the biggest shit-hole place I've ever been.

The main street was shops, banks and bars. The place was still waking up as we wandered along looking in shop windows. Away from the main street it was pretty much a shanty town. A small, flatbed truck had stopped on the sidewalk outside a bar on the corner of the street. On the back was a huge block of ice about four feet cubed.

The Mexican driving the truck got out, fag in mouth, and climbed up behind the ice.

Standing proudly in his filthy white vest, he began kicking the ice with the flat of his shoe until it slid off the rear of the truck onto the sidewalk outside the bar. A curious dog cocked a leg up on it and squirted down the side of it.

I found out later that it was the ice for the day at the bar. Nice!

The intention of our visit was to have breakfast in Mexico. Just to be able to say it. You know how it is. Well, we were going off the idea. We wanted to change a travellers' cheque and wandered over to the bank. It was nine o'clock. We didn't know what time the banks open so, as you do, we tried the door. It was open. Great. We wandered in, as you do.

Immediately two guards, armed to the teeth, sprang into action. We found ourselves standing in the middle of the floor with two handguns held out at arm's length, pointing straight at us. We looked at each other in utter astonishment as one of the guards shouted something loud at us.

I said, "We want to change a travellers' cheque."

"The bank's not open," said one of the guards, looking as bewildered as us. "The door's open," I told him.

"The bank don't open till ten," was the answer. "Well why's the door open?" I asked.

Still pointing their guns at us, one of them repeated, "The bank don't open till ten."

Trying to keep calm, I said,

"OK, we'll come back at ten," and started wandering to the door, guns following us. We went back outside.

"What was that all about?" asked our Ed. "The bank don't open till ten," I told him. "Yeah, I know," said Ed.

There was some sort of food wagon on the sidewalk selling hot dogs and other shit like that. There was some kind of kebab meat hanging covered in flies. When the vendor wanted to cut some off he swished away the flies, which waited until he was finished cutting then landed like a cloud back upon it.

"Fancy a hot dog?" "Pass."

It was nine thirty as we arrived back at the border. A jolly American with a huge grin examined our passports. He leaned out of his booth to hand us them back.

"It wasn't that bad, was it?" he laughed. "Welcome back to the US of A. Have a nice day."

The first diner we found we stopped for breakfast. Mexico had been a complete washout. We had one more thing on our agenda to do and discussed it over breakfast. Catalina Island. We had read a lot about it and had promised ourselves that, if we had time, we would try to fit it in.

To get there you had to take a helicopter from Long Beach. It wasn't far from San Diego, so we decided to go tomorrow morning. We finished breakfast and decided to spend the rest of the day at the hotel pool.

After a small amount of sweet-talking, the hotel receptionist arranged our booking for the next day. The helicopter left every hour on the hour. With a provisional booking we just had to turn up and pay. We were on the ten o'clock flight.

With that sorted, we decided to eat on the *Reuben E Lee*. It was extravagant, but what the heck? We might never be back this way again.

We left Harbor Island first thing the next day and, by eight o'clock, we were parking the car at Long Beach, where we had the surprise of our life. The *Queen Mary*. We hadn't known it was there. It had been converted into a floating hotel.

The Americans had tried to create a piece of England around it. There were red English telephone boxes and a few period shops. A long walkway led up to it, at the bottom of which was an advertising board for breakfast.

"We've got to check this out," said our Ed.

"Too right," I told him.

We walked up onto the deck of the great old ship and wandered around in awe. I remember it smelled very musty. The carpets were old and worn. It had seen better days and was in need of refurbishment. We followed the signs for the restaurant. We were shown to a table and given menus.

We decided to spoil ourselves and ordered steak, eggs over-easy, toast and a huge pot of steaming coffee.

The waitress was lovely. "You guys from London?" she asked. "Well, England, anyway," I told her.

"Gee, that's great!" she cried. "I don't think we've had any of you English guys here before. You on vacation?"

"Yeah," enthused Ed, talking to her ample cleavage. "We love it here. Everyone is so friendly."

"Sure is easy, with nice guys like you," she smiled. "I'll get your order."

Not able to help himself, Ed launched into his old, corny chat-up line. "Did anyone ever tell you, you've got come-to-bed eyes? But are they come-to-bed Eddy eyes?"

"You guys," she laughed, and walked off.

"I think she fancied me," said Ed, testosterone boiling over.

"Yeah," I grinned with affection for my kid brother. "We're going to miss our flight." "Doesn't matter, we can get the next one."

I knew he was right. We were having an added bonus to our trip. Breakfast on the *Queen Mary*. We never expected that.

An hour late we arrived at the helipad only to be told that all flights for the day had been cancelled. The ten o'clock flight had crashed.

No survivors.

We looked at each other and felt a cold chill. We drove back in almost total silence to Harbor Island. We were both thinking of Phil and how close we had almost come to joining him.

Another timely reminder. Here today, gone tomorrow.

Back at the Ramada we hit the bar and partied. It was almost the end of our adventure.

We reminisced on all the fun we had had and how close we had come to death.

We felt a little melancholy as we left Harbor Island for the last time and our flight home. The only thing left for us now to look forward to was our flight home and a free bar on Pan Am.

We soon had them trained. "I know. Two more beers!" Happy days.

The oldest and the youngest
Grandad Pop annd Daughter Angela

Mom, Dad, Me, Phil and Ed.

Phil. Tragically shot age 17 yrs.

June and Roy from Jersey with brother ED

Me and my Jensen Interceptor in the Seventys

The Queen Mary at Longbeach. Caused us to
miss our helicopter and a date with death.

USS Puget Sound, Ken's Ship

Ken and I on the battleship USS Puget Sound AD38

Me and a young Kimberley

Me Barbara and Kimberley at the Belfry

Me in the water, (centre) after a dive at Stoney Cove.

Me and the Lord Mayor of Birmingham,
raising funds for charity at Highbury Hall

Thirty-four

We were finding it hard to settle back into work after the excitement of America. Ed was back grafting with the Old Man and I was slogging away again in Somerville Road.

I'd been thinking about moving house for some time so, when I got to hear about a really nice 4-bed detached in Penns Lake Road, Walmley, I went after it. It was a prime location and going cheap for some reason. I can't remember why, to be truthful. It was a long time ago.

What I can remember is that it was up for about £55,000. I offered £47,000, telling the agent that's all I could afford, looking at the price I would get for my own. Mortgages weren't that easy to get and I told them that I wasn't really all that bothered one way or the other. The bottom line was that I could not afford any more but, with my good standing at the bank, I could go ahead without selling my own first.

Basically, take it or leave it. They took it.

I was looking to liquefy my assets and sold off the rent houses after the old lady passed away and the other two became empty.

I wanted to take Rita and Angela to California, as I thought it was a great time to see the place. Not wanting to go in the mad tourist season, I booked for two weeks in October. I got just the flights, as I knew all the ropes on car hire and hotels.

It was only February 1980. October seemed a long way off. I was having a drink with Siddy Pickering and he told me that the lads were planning a week away about May. Brian, my old diving buddy, was going. If I went, there would be eight of us altogether. They wanted to go somewhere different.

I suggested Malta. I knew there was some great diving over there and the place had a lot of history and places of interest. Everyone thought it was a great idea, so we left it to Sid to organise things.

He found two flats to let in Sliema that were over a pub. *The Anchor Bar.* They were right on the seafront and were both two-beds with lounge, kitchen and bathroom. They were part of a package deal that included flights and two hire cars.

It seemed no time from thinking about it before it was upon us. We stepped onto the stairs outside the plane and were hit by a wave of red-hot air. The sky was cloudless and, even in just a minute, you could feel the sun burning your skin. We would have to be careful.

Customs and Immigration were no problem. We met our rep and she guided us to the car-hire desk. She told us briefly all about the holiday and showed us on a map where we were going. She said she would be visiting us later at the apartments.

We sorted out the cars and, as I was the most experienced driver, I was tasked with finding our digs and getting us there safely. Me, Sid, Brian and Pete took the first car and Gissle, Stuart, Dave and John followed in the second.

Studying the map, it seemed easy enough. We were in Luqa. If we headed to Valletta and took a left around it, Sliema was not far. We found it and parked up. *The Anchor Inn* was on the coast road overlooking a rocky shore that led down to the sea. The road consisted of shops and bars and small guest-houses that led down to the town square. It all looked very pretty.

The heat made us thirsty and we figured a few pints in the pub was a good way to settle into the holiday. It turned out that the flats were owned by the bar and they had the keys. The natives were friendly and the girls all pretty, with smooth, silky, olive skin. The beer was good and so cheap that you almost didn't notice spending any money.

It was only about four in the afternoon and, wisely, I said it would be best if we got ourselves into the flats and checked out things we would need.

The flats were spacious, one on top of the other over the bar. The trouble was, they were dirty. The fridge, cooker and bathroom were far from acceptable. The whole place was in need of a deep clean.

Nobody was happy. A revolution was brewing. Threats and recriminations were flying around. They were saying they wanted to see the rep and get transferred elsewhere. They weren't staying in this pigsty. All that sort of stuff.

Well, to be fair, I'm used to cleaning other people's mess up. With houses and shops left dirty by other people, I'd had to get stuck in myself many a time. Me, Sid, Brian and Pete bagged the downstairs flat. I looked at my watch.

"Get the hot water going," I said, authoritatively. "We can stand here moaning about it all day, but it's going to get no better. There are mops, buckets and cleaning stuff in the cupboards. Let's get stuck in. The sooner we get it clean, the sooner we can have a pint."

With a real spirit of enthusiasm, we set to it. Crockery in the sink, clean out the cupboards, mop the floors. We cleaned the bathroom and found a vacuum. We went all through the flat.

Bleach, disinfectant, soap and water, you could taste the difference. The fridge was spotless. That was the main thing for me.

Two hours later it looked like new. Well! I mean it was clean. We all stood looking at it, quite proud.

"Well done, lads," I congratulated them. "There's a little shop on the next corner; let's do a bit of shopping. It's Sunday tomorrow, they might not be open."

With a satisfying spring in our step we traipsed off to the shop. Coffee, tea, sugar, milk, bread, sos, bacon, eggs, tins of beans, toms and soup and we were set for the weekend.

Down in the *Anchor*, on the bar, was a glass cabinet with huge, fresh baguettes overhanging with various meats and lettuce and tomatoes. With one of those and a pint of cold, local beer, we found a table and settled in for our first Saturday night in Malta.

The other four lads had found the rep and were giving her a hard time. We let them get on with it. The rep asked us what our flat was like and we told her the truth. It had been a tip, but we fixed it. We weren't interested in the others' problems.

The next morning, we cooked a great breakfast and decided to take a look at the surrounding places. The smell of bacon brought visitors from upstairs.

"Where did you get that?" Dave asked.

"Shop just on the next corner," I told them, as I locked up the flat. "See you later."

We climbed in the car, deciding where to go. We passed the corner shop. It was closed.

Oh well! The coast road seemed the obvious bet. Not easy to get lost on and far more interesting than the internal roads of the island. It was a very barren island.

St Paul's Bay was the first stop, then Mellieha Bay and, finally, the ferry port that took you to Gozo. We would go to Gozo another day. Brian and I wanted to get back. There was an address in Sliema that we wanted to find for the next day.

Aquariggo Scuba Diving Centre, Preluna Beach Club, Tower Road. We guessed that it would probably be closed Sunday afternoon, but it would be easier to find today. Then we could go straight there Monday.

We wanted to hire equipment for a couple of dives. Having effectively only six days on the island, two or three dives were the most we could hope for. As it was to turn out, it would be only one. I lost my dive buddy on the first dive and, with nobody else to dive with, I had to quit, but not before I broke every rule in the book by carrying on with a potentially dangerous dive alone. First rule of diving - never dive alone.

Having found the place and satisfied that we could find it again, we parked the car back at the flat and had a few beers and one of their massive baguettes. The locals told us of a great nightclub in the town, very popular with locals and tourists alike. A young people's place with great music and atmosphere.

The other four lads joined us and told us about their day. They had, apparently, had a row with the holiday rep and got nowhere so they, like us, had to make the most of it and clean the place.

After a rest and change we all met at the *Anchor* and had a few beers, intending to go to the nightclub later. About ten, we piled into the cars and found the place. It was, as the locals had said, great. The music was all up-to-date disco. Boney M seemed really popular. The strobe lighting was fantastic and there were two busy bars either side of the dance floor. We all split up and wandered. Eight blokes together can seem a bit intimidating. We all had our own agendas.

Myself and Brian liked to talk diving and drink plenty of beer while talking. There were no bar stools available so we stood close by, hoping that somebody would leave. As I've said, Brian had his long, black hair and beard and was quite a handsome bloke. I was more the clean-cut type. Together, we complemented each other and were definitely hoping to pull a couple of birds.

There was a girl dancing on her own who stood out from the crowd.

She was a bit over twenty, I thought. She was wearing a nice, white dress, belted at her slender waist, and obviously British and well-tanned. Her long, black hair thrashed around her face in the white strobe lights as she gyrated to Boney's *Ra Ra Rasputin*.

She looked a tease. She was sexy and knew it. Boys asked her to dance. She didn't ignore them, but just let them jump all around her. It all seemed far too energetic for me. I was closing in on my thirty-second birthday in August and had no intention of jumping up and down on some dance floor like a demented prat. I was enjoying watching, though.

So was Brian, it seemed. He had this annoying habit of flashing his huge, infectious smile of perfectly straight, white teeth from under his mop of hair and bearded face. At six feet three inches, with his fit, masculine body and rock-star looks, I found him at times quite irritating.

And this was such a time.

For some reason I couldn't quite understand, he seemed to have caught the disco diva's eye. She seemed to be bouncing a little higher and more energetically. Boney M finished and, seeming to forget the song was ended, she continued on dancing right over to us.

"Great moves," Brian complimented her. She flicked her hair, right into my face, and gasped.

"This place rocks, it's just so cool. It saps the hell out of you. But I love it."

Brian bought a round of drinks and all three of us got talking. As soon as she had spoken, I knew that she was from our area. It happens that she came from Birmingham, not too far from us. Castle Bromwich. She was lively and great fun. She was dead keen when we told her we were here for some scuba diving.

By the end of the evening, it was going to be a photo finish to see who, if any of us, was going to get a pot at it. Fair play to the girl, she obviously had a split decision on her hands and, as she couldn't make her mind up, decided to call it a draw. It wasn't three in a bed, she decided. It was none in a bed.

She was on holiday with her mom and dad and, when we offered to take her with us for our first dive, she jumped at the chance. We told her our plan for the next day; she loved it and was even excited about it. We explained that we were getting up early, with the hope of being at the dive shop for opening time at 9.00 am. We would hire all the gear, load it in the car and head for the Blue Grotto, the dive site. We would pick her up about 10.00 am. Be ready. She would be. We picked her up.

It seemed she was still very much up for grabs. The race was on.

• • • • • • • • • •

Christ, I thought, as she got in the car smelling of fragrant shower gel; she squeezed her body into the back of the car behind Brian. A black-and-yellow bikini showed off a firm, fit, well-tanned body covered only with a white, see-through lace top that stopped mid-thigh. I'd never thought of sandals as sexy until now. Full of it, I told her all about the Blue Grotto.

A series of underwater caves with a maximum depth of 100ft, they have shore access. The first of the caves is 76ft deep. It has a large number of cardinalfish. The bigger cave (often called the Bell Cave, or Chimney Cave) is just around the corner at the same depth. It has three small entrances, one of which is fairly easy to negotiate. The interior walls are covered in sponges, tube worms, Bryozoans and a few peacock worms. Marine life over the algae beds includes cuttlefish and many species of wrasse. The caves are visited from the sea by small boats carrying sight-seeing tourists. There can be a great deal of boat traffic overhead and great care needs to be taken.

Although fairly sheltered from the northeast winds, they can be blown out after heavy rainfall. The surface cave area sparkles all over the walls with blue crystals. The caves are dark and the walls black. When the boats are inside, the sun shining in through the mouth of the caves heightens the blue walls.

Hence the Blue Grotto.

We found our place of entry and parked the car as close to the rocky shore as possible.

There was a large, flat, smooth slab of rock about the size of half a tennis court that we could use to get changed on. The sea came up to about a foot below it and, looking down into the water, it looked about sixty feet deep.

Although it was blisteringly hot on the surface, I knew that the deep water in the caves could be really cold. It was a right ba lamb squeezing into our wet suits and dive gear. We would need a lot of lead on to take us down to the bottom. I was sweating buckets and couldn't wait to get into the water. Brian was pissing about and fumbling with his hoses and mask. It had been a while since he had had a dive. And that was not a sea dive, it was Stoney Cove.

I was showing my concern as I said to him, "You all right, Bri? You look a bit flustered."

"I'll be all right when I get in this sodding water. The heat's killing me," he snapped.

The disco diva seemed fascinated by it all. She'd laid out her towel on the rocks ready to top up her tan while we were in the water. She was also under strict instructions that, if we weren't back in an hour, then there was something wrong and she was to go and get some help.

The truth was, though, if we didn't come back, in all probability there was going to be nothing anyone could do for us anyway. It was quite sheltered where we were and the dive plan was that we had to dive to the bottom, then swim out seaward keeping the rocks on our left. We had to swim out into the open sea and follow the rocks around for as long as it took to get into the caves.

All sounded very simple.

The disco diva's name was Liz. We did a full check of each other's equipment. Brian's mask was steamed up and, inside, I saw a look I didn't care for. Fear.

We gave the thumb to forefinger sign for OK? And stepped off the ledge into the mercifully cooling water. Another OK? sign and we began slowly sinking, feet-first, down the side of the smooth rock-face. The water was clear, with good vis. Almost together, we touched bottom at sixty feet. I turned to Brian; he was wrestling with his mask. It was full of water and he was struggling to clear it.

For Christ's sake, I thought. It was simple basic training to clear your mask. You lift the bottom away from your face, tilt back your head and blow air from your nose into the mask, forcing out the water, then jam it back onto your face.

He couldn't do it and gave me the up signal and took off to the surface. Shit! I had to follow him. He surfaced a few seconds before me. Liz sat up sharply, her ample, naked breasts bobbing on her chest, and pushed her sunshades to the top of her head. She said,

"That was quick."

"What's the problem?" I asked Brian, as I spat out my regulator. "Can't clear my mask. Must be hair jammed under it," he coughed. "Clear it up here," I told him, "Then go down carefully."

"It's no good. I'm not doing it," he was gasping.

"You what?!" I exclaimed angrily. "After all the trouble we've gone to?"

"Can't do it, Des, I'm sorry. I'm out of practice. I can't do a sea dive. I'm knackered." He looked genuinely upset. "Sorry, Des."

"Don't worry, mate," I consoled him. "I understand. It's a big ask after a lay-off. Just one of those things."

He creased his mouth into a frown and, looking completely dejected, he said again, "Sorry, Des."

"Don't worry about it."

Honestly speaking, it was far better to happen now than out in the sea. If he'd panicked out there, God knows what could have happened. However, I hadn't gone through all of this to just abandon the dive before it began.

"OK," I said. "Keep an eye on Liz for me."

I could see that that wasn't going to be a problem. "See you later."

I jammed the regulator back between my teeth and sank back down the rock face to where I was ten minutes earlier. I checked the time and my air. I had lost a lot of both. I touched the ocean floor with a small cloud of sand and kicked out to the open sea, keeping close to the wall of rock on my left.

As I neared the open water I was buffeted lightly by the heavier sea as it crashed, not too heavily, against the rocks above. Swimming on the bottom, I made good time and, aware of my loneliness and vulnerability, I was glad to see the mouth of a large cave emerging from the cloud of sand ahead.

I finned into the gaping mouth of black rock. The sun fired arrows and lances of coloured light down through the crystal water. Everything was as I described it earlier. The fish life was wonderfully colourful and varied. I decided to go up for a better look. I checked my depth. Sixty feet. I checked my air. I needed to conserve it. Brian's earlier mishap had cost me dear. At this depth I wouldn't have enough air to get me back. I had to go up just to save air. The deeper you are, the more you use.

The lead I needed to get down, I now needed to lose to get up. I slightly inflated my life-jacket, just enough to make me lighter, and began rising slowly to the surface. It got brighter and brighter as I followed my stream of bubbles upwards and my head popped out of the water.

I was in the middle of the sparkling-blue cave. I was bombarded by flashlights going off all around me and realised that there were two small boats full of tourists taking photographs of me. I waved as they called out to me. I hadn't air to spare socialising and guessed that, if I swam back just below the surface at about twenty feet I should, with luck, make it. The last thing I wanted was to run out of air. If that happened, I would have to surface and fin back on my back with my life-jacket inflated. In a choppy sea by rocks, that would be hard work and take a lot more time and energy than I wanted to spend on it. I wanted a beer.

And that's how it was. I made it comfortably back to the exit point. I spat out my reg, tore off my mask and dragged back my steaming hood and gulped in some fresh, warm air.

Little miss big-tits bounced up onto her bum, shouting,

"How was it, Des? Will you tell me all about it? I was so worried about you. Are you all right?"

"I'm fine," I called. "A few scary moments, but it was so worth it. I've got to be honest, though, I think I shit myself a bit when those three sharks started hovering in the mouth of the cave, outside."

"Sharks!" she screamed. "Wow!"

"I don't think they were man-eaters," I said, trying to calm her fears. "They were pretty big, though. Give us a hand out, Bri. I'm out of energy."

Liz rushed to my assistance, hanging over me with two hands helping me onto the ledge.

Awkwardly, her swinging breasts kept getting in the way of my face. I suppose I could have managed on my own. I knew she was only trying to help so I put up with it, not to offend her.

Happy days.

• • • • • • • • •

Seeing that our diving holiday was obviously over, I decided to take the diving stuff straight back to save pissing about. It was about three in the afternoon when we got back to the *Anchor* for a baguette and a beer. Liz wanted to know every detail of the dive. She was all over me like bees on honey. Brian seemed pretty subdued. I felt sorry for him. He'd lost a lot of face and I tried not to say too much.

He'd had a panic attack and lost his nerve. It could happen to anyone. I said to Liz,

"Liz, don't think I'm being funny, but Brian's feeling a bit down over what happened and I don't want to talk too much about it while he's around. You understand?"

"Of course."

"Look I've got an idea," I said. "What about if I pick you up this evening from your hotel and we have a few drinks here? I'll prepare some food and I'll cook a nice meal for us in the apartment and I can tell you all about it. I don't really fancy another night in that nightclub. You just can't hold a proper conversation. You're a beautiful girl and there's so much I'd really like to learn about you. I find you so interesting to talk to. What do you think?"

"Des, that sounds fantastic to me," she said, enthusiastically. "I'm loving it. I only go clubbing because there's nothing else to do other than sit with Mom and Dad all night in the hotel."

"Well I'm so glad you said that," I told her, with great relief. "It's really taken it out of me, that dive. I'm quite shattered. I broke every rule in the book today, diving alone. And those sharks!"

"Shush," she scolded me. "Save it for tonight." Leaning into my ear, she whispered, "I've got a little secret of my own. I never wear knickers. Never have done."

Quite why she told me that I was never sure, but it was all the incentive I needed to make sure I cooked up the dinner of my life.

"Is there anything I can do to help with the dinner?" she asked. "Yes," I said, "You can take care of the sweet."

"Toad in the Hole all right?"

I told the lads about my quiet evening. Brian was not happy.

"You going to that club again?" I asked. "It was great in there. I don't suppose you'll be back till late."

"Do we have much choice?" Sid asked. "I wondered how long it would take you." "What?!" I protested.

"Don't 'What?' me, you prat," said Sid.

"Haven't got time to talk," I said, "Got shopping to do. See you later."

I did a supermarket sweep for a bag of goodies that could make up a simple meal. Garlic prawns to start. Steak, chips, peas, beer and wine. Ice cream and chocolate for dessert.

Liz looked sensational when I picked her up from the hotel. She had obviously made an effort and wore a lovely, light summer dress and little cardigan. If she was telling the truth, nothing else. I took her back to the *Anchor* and parked. We had a few beers and vodkas with the lads and then went up to the flat to eat.

With plenty of wine flowing, I showed off my cooking skills with flaming king prawns in garlic, followed by steak, chips and peas smothered in black pepper sauce. Over ice cream and chocolate, I told her the thing I hated the most about diving. Shark attacks. I confessed that when I saw those three great whites hovering outside the cave that morning, with me trapped inside, I was scared.

She understood and said, sympathetically,

"Anybody would, Des. It's nothing to be ashamed of."

"Yes, but you don't understand," I stressed. "I didn't have my knife with me." Liz leaned back in her chair.

"That was so good, Des. I'm stuffed. Do you fancy a lie-down for a bit?" "Whatever!"

It was time for truth or forfeit.

The truth prevailed.

• • • • • • • • •

Time flies when you're having a good time, they say. And it did. Liz and I had a couple of fun times together. Nothing serious, just holiday stuff. The lads and I explored as much of the island as we could. Somehow, and don't ask me how, because I still can't believe it myself, we were chased through the streets of the town by some nutcase firing a gun at us.

Yes. You heard me right. I was driving in front, with the second car of lads behind. We stopped at traffic lights and suddenly, behind, I heard the lads swearing and shouting, waving their arms around giving V-signs to somebody. The next thing I knew, there's some bloke waving a gun out of the window of his car, firing shots. The lads behind screamed off in their car with this local racing after them. I took off after them, with Sid, Brian and Pete screaming in my ear. Suddenly I thought, what the hell was I doing? I'm chasing after a man in a car with a gun. I pulled over, turned around, and headed as fast as I could back to our place.

Scared shitless, the four in the second car arrived back later, demanding to know why we had abandoned them. It turned out to be road rage that got out of hand. Well, it was nothing to do with me, and I told them straight that I was not getting shot for anybody.

I still have vivid memories of someone who got shot.

I was glad to get home again to England.

Thirty-five

It was only about three or four months until we were due for our family trip of a lifetime to California. Unlike my trip with Ed, we weren't planning to drive all over the place. I decided to base ourselves at Harbor Island and booked the Travelodge for the two weeks. From there, we could do all the attractions and visit the best places I had found with Ed.

We flew from Heathrow to Los Angeles and hired another Ford Fiesta. It was only a short drive to San Diego and our hotel. Over the two weeks, we did the tour of Hollywood with a map showing us where all the film stars lived, and saw Peter Falk, i.e. Columbo, washing his car on his drive.

We did Universal Studios and SeaWorld and, of course, Disneyland. It was on the way there that I was stopped by the Sheriff for doing 95 in a 55 area. He gave me a citation and I was supposed to post bail at $130 at the courthouse in Vista. I never did and subsequently received threatening letters, which I ignored, resulting in the issuing of a warrant for my arrest.

We had days out to the best beaches, La Jolla and Mission Bay. Restaurant-wise, we favoured the Rueben E Lee, where we saw the Supremes eating. It was probably our best-ever family holiday. Angela had a small, white teddy bear that she'd had from when she was a baby and took it everywhere with her. It was on the bed in our room when we went out one day. Upon our return, it was gone. We turned the room upside down looking for it. Eventually, we came to the conclusion that the maids must have accidentally taken it with the washing when they were changing the beds.

The management were fantastic and searched through hundreds of sheets in the laundry and, after triumphantly finding it, they placed it back on Angela's bed for us to find. It was a great relief, as it was priceless and could really have spoiled the holiday.

Back home again, it was time to take stock of our situation. I had already decided that I wanted to make a little extra money, and the only way to do that was to sell more meat. The shop I was in was at its maximum potential, so the obvious thing was to find another shop.

A shop came up that I thought had great potential. It was in a parade, with a supermarket two doors away and a greengrocer next door. There was a large, busy pub, *The Broadway*, on the corner and a bus stop outside the door. It had everything going for it.

I bought it. It was well fitted-out and looked clean and modern. I took out a half-page ad in the local *Birmingham Mail*, with a nice photo of me and mechanical-electric model butcher chopping meat on a block. Stanley.

Stanley was to become a local celebrity, popular with the kids, who dragged their mothers to the window to watch him, at the same time tempting the moms to buy meat. I decided I wanted to run that shop myself, so I put a manager in the other.

In those days you could get school-leavers on work experience for six months at a time. If they were any good, you could keep them on. Most were a waste of space but were OK for free labour. I got lucky with a couple and kept them on. Leslie Dryhurst was a good, honest, hard worker and learned quickly. I kept him on at the old shop in Somerville Road.

Pete Bassett was another sharp kid and a fast learner. He was a really smart, good-looking lad with ambition. He loved the girls and they loved him too. We became good friends and used to go for a drink quite often after work. He was a bit of a cheeky scoundrel and, if I'm honest, I saw in him a lot of me when I was younger.

Also, he looked up to me and was eager to learn whatever he could from me. You had to watch him, though; he would fiddle a few quid whenever he could. I knew what he was up to and, as long as he didn't get greedy, I was OK with that. He was being subsidised by the government, so it was just topping-up his wage.

The shops had done well over Xmas and, on the Boxing Day, I decided to go over to Small Heath and have a couple of pints with some of the customers. I was very careful, as I didn't want to get breathalysed again. I left the pub early to get back home and walk to the local.

I was in the Jensen Interceptor, driving very carefully, when a copper on the beat walked out in front of me with his arms in the air. I had to stop. He asked me to wind down the window.

He asked if I had been drinking. I told him a couple of pints. He got on his radio and told me he was calling for a Panda car to come and breathalyse me.

I was panic-stricken. I couldn't believe it. All sorts of crazy things were going through my head. Why had I stopped? I should have fled around him. I thought about doing a runner but figured that it would only make things worse.

We stood around the car for ages with no sign of a Panda car. I told the young cop that I was desperate for a leak and couldn't wait any longer. We were stopped in St Margaret's Road, on a small, high-rise council estate. There were a couple of blocks of flats not too far down the road off where we were parked. I told the copper I had to go and started walking slowly towards the two blocks. He seemed a bit unsure what to do and was back on his radio. I'd only gone about a couple of hundred yards when he came running after me. As he stopped me the Panda car arrived.

They breathalysed me where I was and it showed positive. They told me they were taking me to Stechford nick down the road. In the station I took another breath test, which also proved positive and I was placed in the cells. I requested a blood test, so they had to call out a doctor.

A blood sample was taken and one given to me for analysis. I called a taxi and arrived home late for dinner and really pissed-off. Rita was not happy, either. It was a miserable Boxing night.

I knew I was in the shit and figured all I could do was soften the impact and get as organised as I could to get through it. I received a summons to appear in court in January. I took my old drinking-pal solicitor with me, Howard Joy.

I told him I needed an adjournment. He stood before the magistrates and got me a month.

I used that month to try and find a butcher who would buy my new shop at the Broadway.

These things don't happen overnight, but I was in negotiations with a great bloke who everybody knew in the trade, Pat Cody. A Liverpudlian comedy act and first-class butcher.

February came, and Howard Joy came up with some sob story about me and we got another month's adjournment.

To cut a long story short, Pat Cody couldn't get the funds to buy the shop at the time. We made a shake-of-the-hand agreement that he would rent the shop from me and, at a later day, he would buy it at the same price I was asking now. That meant that, with the ideas he had for increasing turnover, I wouldn't be greedy and try to screw him for more money. In those days, trust and a handshake meant more to us than paper. We both were well-respected in the trade and everyone in it knew our arrangement. We both knew the deal was sealed.

In the meantime, Howard Joy had exhausted the patience of the court. In total, he managed to get adjournments in March, April, May and June.

July was D-Day, they were standing for it no longer. They told me to make a plea that day.

I did.

I pleaded not guilty.

The prosecution were stunned. They had never in the reign of Pigs Pudding expected me to plead not guilty.

They asked for an adjournment.

I gave them a big smile.

● ● ● ● ● ● ● ● ●

Pat Cody had officially taken over the shop. He was paying me a fair rent and trying out all his deals on deep-freeze meat packages, which was the big thing in those days.

The Somerville Road shop was running smoothly and it was a great summer. The prosecution had been in touch with Howard Joy and decided that, as I had such a vigorous defence, they would need time to build their case. They believed the court would be needed for half a day and the trial was set for early February the next year.

Thirteen months after the offence!

A lot of people thought I was mad dragging it out. Me, I figured I was going to have to do the time, so I might as well do it when it best suited me.

Having bought the time, I decided to stash as much cash as I could for the rainy days to come. Pat Cody was reluctant to commit to staff until he was certain how things were going. He kept Pete Bassett on and I did some time working for Pat part-time on the cheap. I could still drive and had my van, so I could collect Pat's meat from the market when I collected the meat for Somerville Road. I worked until lunchtime, then went home to Penns Lake Road and sunbathed in the garden with a few cans.

I was booked to go on holiday with Rita and Angela to Spain. The closer it got, the more I realised I needed to be here, not Spain. I was doing a lot of wheeling and dealing at the time, that was a bit too lucrative to leave. Securing my future was more important than two weeks in Spain.

However, being the man that I am, I didn't want my family to suffer because of my sins.

Even though it was impossible for me to go, I still insisted that they went.

I managed to get a full refund for myself from the holiday insurance, based on the rapid deterioration of my mental health. It was only a couple of months later that things would finally catch up with me and I ended up with a full-blown nervous breakdown and rehabilitation in *Uncle Willies Bar*, Torremolinos.

I reluctantly waved Rita and Ang off on holiday, after they promised to phone me regularly to let me know that they were all right.

The next day, Sunday, I met my friend, Gary Fitsgerald, who I knew from the meat trade and the *Bagot Arms* pub on the Chester Road. We were in the *Bagot* with a lot of the local lads, having our usual lock-in. I had left the car at home, obviously, so, when we'd had enough and it was time to go, Gary offered me a lift.

I had a bar built at home and it had seen a good few parties. Gary liked a vodka and, as we were both single, so to speak, I invited him in for a tot. We were getting well through a bottle when I needed the toilet. It was upstairs. The stairs came off the bar and had a large, wooden, ship's steering-wheel fixed to it at the bottom. It worked and spun well with a push.

Fully in the party mood, I gave it a good spin as I started up the stairs. However, in doing so I lost my balance and crashed onto the wooden upright of the stairs. Cursing with pain, I continued up to the toilet. A few more vodkas later and Gary decided he'd better go home. I saw him off and went to bed. I didn't wake until the next morning.

I was in agony. Never known such pain. I could barely get up. I managed to get dressed and knew there was nothing for it but the hospital. I got downstairs and managed to get into my coat. I closed the front door behind me and stood in the porch to get my keys out for the porch door and the van. No bloody keys. Shit. I bowed my head in disbelief. I was locked in the porch.

The front door was glass. A big piece at the top, a smaller square at the bottom. There was nothing I could do. I smashed in the bottom pane and, in total agony, crawled through into the house. They keys were in the kitchen. I got them and let myself out, locking up behind me.

I drove the van to A&E at Good Hope Hospital. After a while, they examined me and told me I had two broken ribs and that there was nothing they could do for me. They offered me painkillers or, they said, I could go home and have some more of what caused the problem in the first place. I chose the latter and phoned the shops to tell them not to expect to see me that day.

I had to do the market Tuesday, so I phoned the shops and got a list of what they needed. The lads at the market took the piss, obviously, but did their best loading the van for me. I went to Somerville Road first and Trevor, the manager, along with Les, the work-experience kid, unloaded their share.

I was hobbling around, holding my ribs, when one of my customers' daughters came in. She was on two walking-sticks, with her feet and ankle bandaged. We got talking about our injuries. Susan was about twenty-two or three. Her mother was Polish and lived not far from the shop in St Benedict's Road.

She was quite beautiful, slim, natural blonde, and told me how she had been in the South of France when she had stood on a sea urchin. Unfortunately, a few of the spines had broken off and she had got an infection.

I told her all about my diving experiences and that I had actually eaten sea urchins. You cut them open and there's a small orange piece in the middle that is quite a delicacy in France. I told her I'd broken my ribs and she thought it was quite funny.

I told her I had to get off to the other shop and said I hoped to see her around. I delivered the meat to Pat Cody and, try as hard as I could, I found I couldn't do much. I told him I was going home. It was another lovely, sunny day, so I spent it in the garden.

No market the next day, so I went straight to Pat Cody's. He was paying me, after all! Not lifting anything heavy, I managed a few hours of light duties. I was in the back when Pat stuck his head through the door.

"Des," he said, in his large Liverpool accent, "There's a nice lady asking for you."

I looked into the shop. It was Susan.

"Send her through," I said.

Pat laughed.

"I thought you might say that."

Susan had been to the doctors not far away and seen the van parked by the shop. She called in to see how I was. I told her I was in a lot of pain and had done a few hours but couldn't take any more. I was going home.

She seemed genuinely concerned to see me in such exaggerated agony. She asked who was looking after me. I confessed that I had no one. I told her how Rita and Angela had had to go to Spain without me because of all the trouble I was in.

I said I was going to lie in the garden and try to get some sun and rest. She told me how she was doing nothing either and, if I liked, she would come with me and try to look after me. I told her how I couldn't possibly put her to all that trouble, but I was in such pain! However, if she didn't mind, I'd be very grateful.

She would hear none of it. It was no trouble at all and she insisted. Reluctantly I agreed, and I told Pat that Susan was taking me home to nurse me for the afternoon.

"I don't blame you, our kid," he told me sympathetically. "Take it easy."

"I'll see you tomorrow," I gasped slightly, clutching at my ribs and wincing with a wink. "Phone you later for the market."

I think I overplayed things a bit for Pat's benefit. After all, Susan was just a friend, injured like myself, and at a loose end on a beautiful summer day. The house was in a very beautiful, sought-after area of Walmley. It was on a bend and had a huge frontage, with a single drive on one side and double drive on the other. It had a huge double garage, with my Jensen one side and my new, yellow speedboat in the other side. Behind was a sunbed room with a salon-strength sunbed for the winter. I gave Susan the tour.

She loved the bar and couldn't wait to get behind it and explore all the bottles and bric-a-brac. She picked up an unopened bottle of Cointreau.

"God, Des," she gasped, "I love this. Got any lemonade?"

"I've got everything," I told her, which was true. I had a fridge and freezer in the garage stocked with everything you could want. "I'll get some ice."

I left her for a minute and returned with an ice bucket and a bottle of chilled lemonade. Selecting two tall glasses, I mixed two Cointreau and lemonades. The doors in the bar opened onto the patio. We went outside and sat in the sun with our drinks.

"You've got a fabulous house, Des," Susan said. "It must have cost you a fortune." "Not really," I told her truthfully, "I never buy anything unless it's cheap, and this was going at the right money."

The drink was almost empty, I noticed and asked, "Another one?"

"Please," she said eagerly. "It's just like being on holiday here. I don't suppose you've got a swimsuit anywhere?"

"I can have a look," I said, "But Rita's probably taken them all on holiday. If not, it's very private here, no one can see you. Actually, there are a lot of advantages to sunbathing nature's way."

"And what would you say were the advantages, then, Des?" she asked, with a sly grin.

I creased my lips and looked thoughtful.

"Well, you'd have somewhere to hang your towel," I laughed.

We both laughed. It was one of those where you manage to stop for a moment, then look at each other and start all over again. I got two more cocktails and went for a rummage in Rita's drawers. I came back with a bikini.

"I don't know if this is any good," I told her. "It's the only one I could find. Try it on. Use the bar."

I stepped aside to let her in. I turned my back and half-closed the glass door until she was in the reflection.

"Don't turn around," she warned me.

"Don't worry, I won't. Will you be all right with your foot? If you need any help, give me a shout."

"I'll be OK," she said, as I watched her struggling with her panties in the glass, then, "The top's too small."

"I'm not being funny," I said honestly, "But you've been in the South of France for two weeks. I bet you never wore a top there."

"This isn't the South of France, Des," she said, defending herself.

"We'll I've been in Spain, God knows how many times, and nobody wears a top there, either. I can't see the bloody difference. I don't understand."

I could see they were far too big for Angela's old top. Of course, I knew that when I gave it to her.

"Can I just turn around and get it over with? Then we can relax." "Oh, all right," Susan relented.

I turned around and said, with great admiration,

"Blimey, Sue, you look fabulous. I can't imagine what you were worrying about." "Des, you're my mom's butcher, for God's sake."

"Well, I won't be saying anything, that's for sure," I told her. "I'm not crazy."

We drank and laughed in the sun, with music from the bar in the background. It was a great afternoon. The time didn't bother us. We began to feel hungry about seven, so I suggested we get changed and go out for a meal.

There was a nice Chinese restaurant at the top of Penns Lane, on the Wylde Green Road. *The China House*. I called a taxi and had no trouble getting a table. The place was practically empty. We had a marvellous meal, with some great wine. We finished with some Calypso Coffee and called a taxi.

It took me home first. Susan gave me a nice kiss and I paid the driver more than enough to take her home.

"See you again?" I asked.

"Count on it, Des. It was great."

• • • • • • • • • •

I'm disappointed to say that I didn't see Susan again. I think she realised that we were treading on dangerous ground and that our friendship could possibly be misinterpreted. She reappeared after Rita and Angela returned, and continued visiting me in the shop for our friendly chats. She was a nice friend to have, and extremely easy on the eye.

She certainly did no harm to my image. Her mom was a nice lady and a good customer; I always asked about Susan when she came in the shop. She always told me how Susan sent her best and that she was very fond of me. Nice.

My ribs mended and I dossed a lot, never working past midday. I was in the shop helping Pat when a friend of his, Brian, called in to visit. He was a larger-than-life character who couldn't stand still for a minute. He was a wheeler and dealer by trade and there didn't seem to be anything that he couldn't get you.

Cheap. No questions asked. At night he was a bouncer on the nightclub doors in Birmingham. He was stocky and worked out. He had short, curly, ginger hair and a moustache, and thought himself to be a right ladies' man. He had some tragic chat-up lines.

He had recently split with his long-term girlfriend and had sold their house and had cash in his pocket. He didn't want a proper job. He was able to survive buying and selling. Good luck to him, I thought. We got on quite well and soon he was joining us in the pub after work.

He told me that he fancied going to Spain and seeing if there were any opportunities over there. He fancied himself in a bar. I told him that it would probably suit him. I quite fancied the idea myself. I said so. He suggested we go over together and check it out.

I was all for it. However, I wasn't sure how Rita would feel about it. I fancied a month.

Two weeks and you were only getting the feel of the place.

It was at that point I suddenly realised how all the stress and worry of the last few months had begun to affect me. I was depressed. I couldn't think straight. God forbid, I'd even had suicidal thoughts. I felt as if the whole world was caving in around me. I knew then that I was heading for a breakdown.

A serious nervous breakdown.

I confessed all to Rita.

Everything.

I had to get away for a while and clear my mind, before it completely consumed me. I told her about Brian and how he had said he would go with me and take good care of me. I'd be all right with him. Did she mind? She said it was OK. I told her I would let Brian make the arrangements and let her know the details as soon as I knew them myself.

I purchased two return tickets to Malaga for four weeks. Apparently, at this time of the year, that was all that was available at discount prices.

"How long are you going for?" Rita asked. "A week or a fortnight?"

"Couldn't get a package deal," I said, "The only flights available were for four weeks. Had to take it."

Having my best interests at heart, she told me I could go and hoped I'd come back my old self. I was sure I would.

Thirty-six

Because of a shortage of plane seats at that time of the year, Brian and I had to travel on separate flights. I came out on Sunday and Brian on Monday. I landed in Malaga and collected my baggage.

We were going to stay at Torremolinos. It was the closest large resort to the airport. I had no transfer and was reluctant to take a taxi. I didn't know but guessed that it would be expensive. I already had a great tan from my few months in the garden at home. I didn't look your pale, just-arrived, gullible tourist. I followed a crowd of baggage-carrying holidaymakers who were trailing after a smart-looking girl in a uniform. As I suspected, they were going to the coach park. I caught up with their leader, a friendly girl about mid-twenties.

"Excuse me, love," I said, with my biggest smile, "I wonder if you could help me?" "Sure, fire away," she said in a Newcastle accent. "What can I do for you?"

"I'm after a lift into Torremolinos," I told her honestly.

"Cheeky boy," she laughed. "You don't look like you're new on the streets. What's your story?"

"I'm having time off and travelling," I said, half truthfully. "Where have you been?"

"California," I began, "LA, San Jose, San Diego, San Francisco, Las Vegas, Honolulu, Malta."

"OK," she stopped me, "I'm impressed. Slumming it a bit, aren't we?"

"Needs must. It's cheap here and I might get a little work if I need it," I told her. "I don't want to pay for a taxi, and it's a long walk."

"This lot are mostly going roughly that way. Final destination Marbella. I can drop you off somewhere in town. Where are you staying?"

I shrugged. "Don't know. Haven't got that far yet. I want to rent a cheap studio but I'll need a bed for tonight."

"I've got a couple for the *Principe Sol*," she said, checking her clipboard. "It's on the coast road. If you just want cheap and clean, there's a hotel

Tarik on the front not far away. It's mainly used by Spanish and hitch-hikers. It's nice."

"Thanks, Jane," I said, reading the badge on her chest, "I owe you one."

"You'd better give me one, then," she laughed. "I'm one of the reps at the *Principe Sol*.

When you get settled, look me up. I won't be hard to find."

Hardly able to believe my luck, I sat with Jane for half an hour on our way to Torremolinos.

She told me how she was here for the season and hoped to be in the Canary Isles for the winter. It's not as glamorous as it sounds, she assured me. It was long hours and hard work. Sometimes she had airport duty in the middle of the night followed by a welcome party in the morning.

We entered Torremolinos with the sea on the left and the hotels, shops and bars on the right. The *Principe Sol* was a huge hotel whose grounds began on the coast road itself. Jane called to the driver in Spanish and the coach slowly stopped.

"You need to get off here," she told me. "The *Tarik* is a few hundred yards down the road. If you get off at the *Principe Sol* you have a long walk back. Don't forget, you owe me a drink."

"I won't," I said. "Thanks for the ride. I really appreciate it."

I climbed off the coach with my luggage and, as it pulled away, Jane waved. I waved back. I found the *Tarik* easily. It was a lovely, white, ranch-style building that I'm sure was there before any of its neighbours. Interior was vintage, spotless and 100% Spanish. They had a room that was cheap as chips. It was large and bare, with brown furniture and white walls. The bathroom was clean. I liked it.

It was late afternoon and I phoned Brian to tell him where I was. I gave him the hotel phone number and said if I wasn't there, which I probably would not be, to leave a message or I would leave a message letting him know where to find me.

I didn't bother unpacking. I left my case in the room and went for a wander along the seafront in the glorious sunshine. I found a pavement bar with colour photos of the menu pinned to a wall. I sat down outside at a shiny, stainless-steel table and ordered Cervesa La Grande. One pint of freezing-cold beer arrived quickly and, after studying the photos, I ordered steak and chips, not expecting too much.

When it arrived, it was on a large, oval plate. It was a huge slice of lean beef cut thin that looked like topside. It came with chips, peas, onion rings and a wedge of fresh lemon. To my surprise, it was tender and tasty. The only thing I left was the lemon and a third empty beer glass.

Certain that I was fully fortified until the morning, I began some more exploring, walking along the front checking out all the shops with their merchandise piled out all over the pavement. A few bars later, it was about eight in the evening, I saw a welcome sign.

Happy Hour 8.00 - 9.00.

Vince's Bar, a neon light flashed.

There were a few people in there. It was a small, not really memorable place, but the barman greeted me with a large, friendly smile. It was none other than the great man himself. Vince. He asked in a broad cockney accent what my poison was. I told him in my broad Brummie accent, a pint of lager please, mate.

With a deep distrust of all things cockney, I was on my guard. Vince had been here a few years, he told me. I told him my story. I needed a studio flat in town for a month. Unsurprisingly, it was my lucky day.

What about that?

With sleight of hand Paul Daniels would have been proud of, a small, gold business card appeared in his fingers.

Vince Milton Property something or other. Enquiring more details from me, he told me he had a studio in a multi-storey block in the middle of town. It was modern, well-furnished, with bathroom, kitchen and twin beds. In real money, he told me, I could have it for seventy quid a week, cash. For four weeks, even money. "Two and a 'arfton" (two hundred and fifty quid to Brummies).

Seriously speaking, it didn't seem bad. Two fifty all in. No utilities. He said he'd pick me up tomorrow from the *Tarik* and show me. I asked if I could use his phone to ring Brian. I told him to come straight to *Vince's bar* from the airport. We would have either a flat or a room at the *Tarik*. He told me the time he expected to meet me and we left it at that.

Vince collected me in the morning in his small car. He drove me around the streets to a high-rise apartment block; it was just a road off the main route through the town. It was not far from the centre and the main pedestrian area of town. There were layers of steps called The Calla San Miguel that went all the way down to the beach. It was a perfect location. I viewed the flat and decided it was ideal for purpose. It was plenty big enough for two and there was a lift almost to the door. I think it was about five storeys up.

I shook on the deal and Vince moved me in from the *Tarik*. I paid with five fifty-pound Amex travellers' cheques and was well happy with the biggest expense of the holiday out of the way. That left me with 750 quid and 140 to come off Brian.

I'd explained to Vince the expenses and complications of sub-letting. He fully understood.

Brian was well pleased with 70 quid a week.

Happy days.

· · · · · · · · · ·

Brian's taxi dropped him off outside *Vince's bar* about eight the following evening. I was waiting for him and got him a pint. He was moaning like a drain about the cost from the airport to Vince's. I sympathised and agreed - it was, indeed, highway robbery. However, it's only a one-off and you're here now. Don't worry about it. No, I couldn't remember how much I paid. It was in foreign money. You're only here once.

"Vince needs the money for the flat," I told him. "One forty."

He only had travellers' cheques. No worries, Vince could change them. He handed over three fifties. Vince gave him back pesetas. He examined them suspiciously. He didn't like parting with money. When he went to the toilet, Vince handed me the rest of the change.

"Thanks."

We stayed at Vince's until about midnight. He had a partner, Yan, a Scandinavian who spoke immaculate English. He was a nice bloke. They were into everything. Property management was the main thing. Then, if you wanted a car, they could get you one. They sometimes met people interested in purchasing a property. For a small fee they would help them through all the difficulties. And there were many. Secretly, they confided in me, they would never buy anything themselves. It was far easier to rent and let somebody else have all the hassle.

They didn't have a taxi business but often, people who didn't like the crazy Spanish way of driving would ask to be taken somewhere, or even have a day out to another resort. Airport transfers were always popular with people not on package tours. There were also owners who didn't want to pay heavy taxi-fares. Along with that, they would then offer to check on the property while the owner was away. Maintenance, grass-cutting, cleaning, pool care. They could arrange all that with people who wanted to work under the radar.

Pretty sweet, really. There was plenty of work for people like Brian and me if we wanted it. The holiday reps gave small commissions for anyone you got to fill their coaches for excursions. Brian's ears were pricked.

It was about a twenty-minute walk to the flat. We took turns carrying Brian's luggage. He liked the flat and said,

"You done good there, Des."

We had an early night and turned in about one am. The next morning, while Brian was unpacking and settling in, I offered to go to the local supermercado. I had suggested a kitty. We put twenty quid each in and any change went in a jug that we would top us as needed.

I came back with a trolley I had borrowed from the owner and went up in the lift to the flat.

In the lift with me were two sisters. They were in Flat 53, they said. Their parents owned it. They came from London and spoke very well. Their father was a bank manager and they were schoolteachers.

The lift stopped at my floor and, as the door opened, I pushed the trolley half out. I told them we had just arrived and would be here for a month. It would be nice, perhaps, if we could get together some time so they could show us the ropes. I explained how we hadn't been around very much and that we were concerned about getting taken in by the local natives. You hear so many things about these foreign countries. They told me not to worry. The people were nice. They had been coming here for years and knew lots of nice, friendly places to go.

They offered to show us around on one of the nights. I thanked them and told them if they ever fancied a drink and we were in, they would be very welcome.

The lift door closed behind me and I pushed our supplies to the flat. I told Brian about the teachers and how I'd played the helpless Englishman abroad carrot.

"You silly prat," he blustered, "They'll think we're a right pair of wallies."

"Trust me, Bri," I said, "They're a couple of tasty birds. Two sisters, both teachers. They need a project. Someone to save. A couple of mother hens."

"Yeah," he scoffed. "I suppose you've told them you're still breast-feeding." "Very good for you, Bri. Very sweet, breast milk."

"I wouldn't know," he said, inspecting the shopping. "You spent forty quid on this lot?" "Not all of it," I protested, "There's change. It goes in the tin."

"Where's the food?"

"It's in the trolley, you dickhead."

"There's bleedin' wine, beer, cider, gin, vodka…"

"Yeah! And what's that? Bread, milk, tea, coffee, sugar, butter, spuds."

"There's hardly anything to eat."

"Well, if you don't like it, do your own bleedin' shopping." "I bloody well will."

"Have we just had our first row?" I asked. "'cause if this is what it's going to be like, I want a divorce."

We both laughed at that.

"Come on, you tosser," he grinned. "Let's have a recce and try not to get mugged by the local pussies."

• • • • • • • • •

It is hard to believe that, only 50 years ago, Torremolinos was a tiny coastal village with one main street and only a score of houses. Since the advent of organised tourism to the coast, unprecedented expansion has converted this small fishing community into Spain's ultimate package holiday destination, and a household name in a dozen countries.

Torremolinos derives its name from the old watchtower that still stands sentinel above Bajondillo beach, and which once protected the numerous water-fed flour mills ("torre" means tower, "molinos" means mills). The excellent climate, nearby airport and superb beaches made Torremolinos the perfect the site for mass tourism. Uninhibited development during the 1960s created a concrete jungle that was designed to fulfil every tourist's requirements, at a price that anyone could afford. Countless bars, pubs, restaurants and discotheques offer everything the young and active could want.

And that was what we wanted!

Daytime activities centre on the large, sandy beaches, while the night life is the busiest along the Costa del Sol. Torremolinos is split into three principal areas. The town centre contains the major shopping zones, restaurants and evening bars. El Bajondillo and Playamar, to the east along the seafront, accommodate most of the big hotels. La Carihuela and Montemar, to the west, offer more large, beach front hotels, fish restaurants and a wide choice of nightlife.

Bajondillo beach suited us for today. We made our way down the steps of the Calla San Miguel, through the town. I noted the old Spanish guy barbecuing whole chickens on an open, revolving spit. They were about a quid each! I thought, "I'll be having some of that later."

At the bottom of the steps was the long promenade of bars and shops. The beach was wide, miles of long, hot, golden sand that burned your feet. Every few hundred yards there was a beach bar with a barbecue and sunbeds. We took a table in the sand and were quickly served with freezing pints of beer.

"What we doing tonight?" Brian asked.

"I've heard a lot about *Uncle Willie's* Bar on the front," I told him, taking a long draw on my pint. "It's one of the liveliest places outside of the town centre. Vince's place is all right for a couple, but you couldn't spend the night there. I say we check out *Uncle Willie's*, then work our way back towards the flat, exploring the town centre on the way. And *Uncle Willie's* is only a few hundred yards from the *Principe Sol*, where that rep works."

"What rep was that?" Brian asked, sounding surprised. "I met her at the airport; didn't I tell you?"

"Don't think so."

"Yeah. She gave me a lift into town on her coach. Jane. Jane was her name. Really nice girl. The taxis cost a fortune at the airport. She said I owed her a drink and she sometimes goes to *Uncle Willie's*."

"I thought you got a taxi from the airport," Brian said. "You told me they cost the earth."

"No, you said they cost a fortune," I corrected him. "I just agreed with you. I didn't say I took one. Jane said they cost far too much money and offered me a ride."

"So, you never paid for a taxi!"

"That's what I just said, isn't it? Jane gave me a lift. Anyway, what difference does it make to you?"

He was sulking now. I wish I hadn't said anything. Well just a bit.

Not willing to pay for sunbeds, we laid our towels on the sand not far from the bar. It was a cloudless sky, with a temperature well into the eighties. I carried all my bits and pieces in a small, blue canvas bag with the Scuba Pro emblem on it. It was a round, white circle with a diver swimming in the middle. I took it everywhere with me when travelling. You could always guarantee that it would draw attention from people. A fellow diver would recognise you immediately, or some inquisitive person would ask about it.

Hopefully young and female. It hardly ever let me down.

"Got any sun cream?" asked Brian.

"No, mate. Don't need it. I've got a year-round suntan."

I noticed how white he was. Apart from his face, neck and lower arms, he was lily-white. He gazed skywards, assessing his chances, and declared, "I don't normally burn. I'll get some on the way back."

"Might be a bit late then, mate," I warned him. "It's pretty fierce." "I'll just have a couple of hours, then cover up."

"I'm not one to nag," I began, "but..."

Cutting me off, he said, "I've been out in hotter than this, I'll be OK."

"Your funeral."

From my little bag I produced a paperback book I'd bought for the trip. I liked to read and always took a few books with me. It was a great way to pass the time. I was not a person who felt he needed to be in other people's company all the time. I was quite happy on my own. As I began to read, I reminded myself that, when I retired from the meat game, I was going to concentrate my energy into writing and see where it got me.

I set the retirement target at fifty. I had no intention of slaving my nuts off until I was sixty-five and dropping dead twelve months later with everything on life's wish-list still before me.

I was determined that, when I got to sixty-five, everything on my wish list was going to be behind me. I didn't see myself diving the oceans of the world, exploring continents and shagging its inhabitants at sixty-five.

Instead, I saw the reality.

Aching knees, new hips, blood pressure, diabetes, flu jabs, a bus pass and frequent visits to the doctors followed by referrals to the hospital.

And as for the shagging, I'd never paid for it in my life and I wasn't about to start now. Grandkids, that's what you want at sixty-five.

I can think of no more tragic sight in life than a rich old man with a teenage girl in tow, and everyone taking the piss behind his back.

And while we're at it, what's with the bald head and white ponytail?

The afternoon just flew. While reading and people-watching and guzzling lager, I salivated over thoughts of a sizzling, spit-roast chicken. One final drink at the bar and I suggested we made our way back. The pavement was like an open-air market. I suggested to Brian that he purchase some sun lotion and after-sun. He said he'd get it tomorrow. OK.

We purchased two whole, almost black, chickens and put on a spurt to get them home. We were starving. The chickens were fantastic. We tore them apart and almost finished them. The leftovers we let cool before putting in the fridge for supper. We crashed on the beds and I was woken a few hours later by the running shower and Brian's moans and groans. He was well cooked. Not a hospital case, but I felt his pain. I'd tried to warn him.

"Got any after-sun?"

"I haven't, mate. Sorry. I told you to get some." "Yeah, yeah, I know."

I could only shrug.

Uncle Willie's was above a block of shops. We climbed two flights of steps to the balcony, which was covered with tables and chairs. The bar covered the upstairs area of the four shops below it. It was about half full, with some bar stools available. We took a couple.

Uncle Willie reminded me of an older Siddy Pickering. He was a comedian and a great host. He was everybody's best friend. The tourists returned there night after night until the end of their holidays.

Every two weeks, planeloads of new best friends arrived. I estimated that, by the end of the season, Uncle Willie must have thousands of best friends. If they all kept their promises to return next year, I figured he would have to buy his own high-rise hotel to sleep them all.

Perhaps that was the plan.

It fascinated me how much trust was involved in being a bar-keeper. Nobody, without exception, paid for a drink until they left. I couldn't for the life of me see that working in Birmingham, or Liverpool, or Newcastle.

After the first week, he wouldn't have a friend in the world!

Anyway, after a couple of hours, his two new, really, really best friends started for home. As planned, we checked out a few bars in the town and got to the lift in the flat at about 2am.

"Hold the lift," came the cry from behind. It was the two teachers from Flat 53.

Apparently, they'd had a good night as well. We all introduced ourselves and, as we rose up in the lift, I asked if they fancied a nightcap. They accepted and, upon entering the apartment, declared that it was an exact replica of their own, even down to the furniture.

We chatted and drank for an hour and I saw them to the door, saying we must have a night out sometime. We all agreed, and I was glad to see my bed after picking clean the carcass of my chicken.

· · · · · · · · · ·

You can't change the habits of a lifetime. The next couple of days we spent on reconnaissance. I, in particular, was there for a month and wanted to see how cheaply you could live. Brian was with me all the way on that one. He was hoping eventually to perhaps spend a lot of time there and was all for my money-saving ideas.

With the accommodation all paid for, it was really just down to food and drink, which were cheap enough anyway. I enjoyed cooking and, with all the time in the world, it was fun to shop at the local market, where there was great meat and fresh produce to choose from. It was nice, after a few hours on the beach, to go back to the apartment and spend a few hours on the balcony with a drink while preparing dinner.

The local Spanish cava was a favourite. It was the equivalent of French champagne. As with the French version, it varied hugely in price. In the local supermercado I found one that seemed very popular in the local bars. It was about one of the cheapest. It cost 195pts a bottle. Less than a quid. You could buy wine - red, white or rosé - in 1-litre boxes, for 120pts. It tasted fine. Packs of beer were virtually for nothing. We stayed clear of spirits.

The spit-roast chickens were irresistible and, with a bottle or two of cava, were great on the balcony in the evening.

Another great tip is the happy hour. Every bar had one. Not all at the same time. Some did

4-6, others did 6-7, 7-8, 8-9, etc. With a little ingenuity, you could drink half-price until ten in the evening. Many of the bars put small bowls of food on the bar, "tapas"; you could snack free while having a beer. Exploring the back streets of the town, there were many small bars that only the locals seemed to use. They were always friendly and welcoming. These were the places we preferred.

Being a bit rough and ready ourselves, we always got on with the locals, many of whom were older and believed in mañyana like us.

"Never do today what you can put off until tomorrow."

We rarely ever left without a complimentary tiny black coffee and a little brandy.

If we felt hungry at lunchtime there was a local cafe, the Spanish equivalent of Joe's Caff, where the "meal of the day" was 295pts. For that you got a plate of hot food. Some sort of meat or chicken. It was always delicious. With that you got a crock jug of red or white wine, to your taste. To follow up there was cake and coffee.

Happy days.

Staggeringly, the most expensive thing in Spain was an English newspaper. It cost roughly the price of two pints or a meal of the day. Needless to say, I never bought one of those. You could always find a discarded one on a café table or poking out of a litter-bin. Hotels were the best source.

Another thing hugely expensive was a paperback book. I get through about four a week, easily, while sunbathing. I found a second-hand bookshop in a back street where they were really cheap. Better than that, though, I found that some hotels had a swap-shop corner where guests placed their finished books and took someone else's. These were a better source, as most of them were the latest best-sellers picked up at the airport.

It was nearly the end of the first week. I decided to look up Jane, the Portland Holidays rep. After an early dinner on the balcony, I showered and changed for the evening and told Brian I would meet him later at *Uncle Willie's*. I took a stroll to the *Principe Sol*. It was one of the largest hotels in town, with 600 rooms.

It was the start of dinner-time and the place was a hive of activity. The herd was stampeding to the dining room. I guessed that they must run out of food after an hour. It made me grateful for the tranquillity of my small apartment.

I found the noticeboard in Reception and found the Portland space. Jane would be there between 8 and 10 to answer any of her guests' queries. I checked my watch. It was 7.30. I found the bar and sat at it. The friendly waiter took my order for a pint and set down two before me. As I looked to query him he announced, "Happy hour, senor, two drinks for the price of one."

"Gracias, amigo." "Da nada."

I passed an idle hour chatting to the barman and nibbling the bowl of nuts he put before me. I love peanuts but do wish they wouldn't give them to me, because I can't leave them alone until they are all gone.

I checked my watch and, after excusing myself to the barman, I wandered to the lounge area. I stood reading the various noticeboards and noted all the excursions on offer. They all seemed very overpriced to me. As I turned away from the board, I noticed Jane in her orange blouse walking across the room, cradling a clipboard.

"Can I interest you in an excursion, sir?" she asked me, in a very professional voice. "Yes," I said mischievously, "But I don't see it on the board."

"My private excursions aren't advertised," she said slyly. "Only for the chosen few, then?" I said.

"The very few," she replied. "Any tickets going?" I asked.

"Don't know yet. You have to be vetted first."

"What time do you finish?" I asked. "I'm meeting my mate at *Uncle Willie's* later." "Unless there's a problem, and up to now it doesn't look like there will be, 10 o'clock.

Half hour's paperwork and I could see you about 11."

"Sounds good," I said, as I watched a few tourists begin to converge on her. "I think you've got customers. I'll leave you to it. See you later."

Brian was already in the bar and bought me a pint. He was talking to Uncle Willie. The bar was filling up. It was Saturday night and the local Spanish Romeos were out on the pull. We watched with amusement as one young tourist girl after another fell for the steaming pile of horseshit that landed before them.

When Jane arrived, I gave her my stool and stood behind her. She was well-known in the bar and a Bacardi and Coke was set down before her. The three of us chatted away.

"Have you scored yet?" Jane asked, laughing. "No, I haven't," I told her.

"Is that why you came to see me?" she teased. "No, it's not," I protested.

"Just kidding," she said, sounding slightly tipsy. "Thought you'd have pulled by now, though."

"I'm not that kind of boy." "Yeah! Like hell."

"It's true, I tell you," I protested further. "I've been saving myself for you all week. And now you're making fun of me. It's not fair."

"Ah, diddums."

We flirted for the rest of the night. The banter was good and Brian was enjoying the crack as much as us.

"Des, I must be going, love. I've got the airport run tomorrow afternoon. Do you think you could see me to my apartment? It's not far and really, it's not safe at this time of night for a girl on her own."

"Yeah, no problem," I told her seriously. "Can you sort the bill, Bri, and we'll square up later?"

"Sure, no problem, Des," he said. "I'll see you back at the flat."

Jane gave Brian a kiss on the cheek and we strolled arm-in-arm down the steps to the street. Jane's flat was a bit further out of town, on a small estate of about six tower-blocks. The endless beach was on my right and I could smell the salt and seaweed as the tiny waves lapped onto the shore. The moon was bright and the light played tricks as it danced on the breakers. The squawks of a few night birds followed us along the promenade. Jane pointed out her flats. They overlooked the sea. When we reached Reception I walked her in to the lift.

"Are you coming up?" she asked. "I think so," I told her honestly.

• • • • • • • • • •

We were into our second week, and I think I may have forgotten to mention that Brian's flight ticket was for two weeks only. It was his last week. Although Brian was great company, he didn't really suit my time agenda. I wanted to live like a local and just blend into life as it really was for somebody living and working in Spain. He was treating it all like a holiday.

I decided that, for the next week, I would go all-out to see that he had a good time, without letting on that, secretly, I couldn't wait to get rid of him and be on my own.

I got back home about ten after having some breakfast with Jane. He seemed a bit edgy and wanted to have a bit more action other than drinking and sunbathing. He made it plain that he wanted his leg over. I hadn't come away with that thought specifically on my mind, even though I wasn't averse to the idea. I told him that we weren't joined at the hip and he was free to go and do whatever he liked. He had a key to the apartment and was a free agent. I could always make myself scarce for a bit if he wanted a bit of privacy. We were having our evening meal on the balcony with a bottle of cava when he came out with it.

"What about those two upstairs?"

"What about them?" I asked, as if I didn't know.

"They seemed up for it," he said. "What about asking them out for the night? They more or less said they fancied it. They know a few good bars. It'd make a change."

"Yeah, that's fine with me," I told him. "Why don't you ask them? They're in 53. Pop upstairs and see what they say."

I knew what was coming next.

"You're better at that sort of thing than me. Will you ask them?" "It was your idea," I teased him.

"But you know them better than me. I only saw them the once," he moaned. "I could tell they liked you."

"Blimey, Bri, I only spoke to them a couple of times in the lift. Why can't you ask them?" I reasoned.

"Go on, Des," he pushed. "I know they'll say yes to you."

Well, I can't say I hadn't already thought about it so, after making him grovel a bit more, I reluctantly relented.

"Oh, go on then," I said. "I'll see if they're in."

I took the lift up to the fifth floor and found 53. I rang the bell and waited. I could smell cooking, so I guessed they were in. The younger sister, Diane, opened the door.

"Des," she exclaimed, "Come in."

Not a bad start, I thought. "Thanks; I'm not interrupting your dinner, I hope?"

"No, no," she beamed. "We're having a few drinks before we go out. Natalie, look who's here."

Her older sister was holding a tall glass that I guessed was Bacardi and Coke.

"Hi, Des," she called over. "What can I get you? We've got beer, wine, or anything, really.

It's a proper little saloon in here."

"Beer would be great," I said. "What's cooking? It smells great."

"Spanish meatballs and rice," she said, reaching into the fridge for a beer. "Want a glass or is the bottle OK?"

"Bottle's good."

"I wish I'd known you were coming, I'd have cooked some extra. Where's Brian?" "Downstairs having a drink. I just had this thought that maybe, if you hadn't any plans, we could go out for a few tots."

Diane said behind me, "You should have come up earlier. We could have all had dinner.

We weren't planning to go out tonight."

"I didn't mean specifically tonight," I said. "I just meant maybe some time, anytime, whenever!"

"To be fair, Des, we were only staying in because we had nothing better to do. There's a great little local bar off the beaten track that we use a lot. If you like, we could go there after our dinner. You'll love it. That is, as long as you don't want some Irish or London bar full of wankers on holiday. It's really Spanish, with Flamenco music and sometimes a dancer."

"It sounds like music to my ears," I answered her truthfully. "I hate all that tourist shit.

Where is it? We'll see you there."

"I think it would be easier if we just call for you on the way out," Natalie piped in. "Save you wandering around looking for it."

I finished my beer. "Fine by me."

She went to the fridge and found me another. "Sit down for a bit."

We talked for a while like we'd known each other for years. They were confident girls. You could tell they were well-educated and from an affluent background. They spoke well, but not posh, and were very easy company. They were no strangers to social circles. They knew how to make you welcome. I liked them and, when I left, I felt confident that we were going to have a good night. They told me what time to expect them and I reported back to Brian.

"Where the bleedin' hell have you been?" he grumbled as I walked in.

"Fixing you a date, what do you think? What's the problem?" "You've been gone ages."

"So what, you ungrateful sod. I've been through hell for you. They're two raving nymphos, those two. I only knocked on the door and they dragged me inside, threw me on the bed and tore all me clothes off. I was lucky to get out alive."

"NA!"

"Course not, you prat. They're calling for us in about an hour."

• • • • • • • • •

The bar they took us to was everything they said. Proper Spanish. We were greeted like family and friends. The staff obviously knew them well. Diane and Natalie had been going there for a few years, ever since their father had bought the apartment.

We found a nice big table and ordered drinks. The owner, Manolo, was about my age, and I got the vibes that he and Diane had some history. He was the typical handsome, over-the-top, Spanish Romeo. I liked him. He was a great host. The evening was a great success. I love all things Spanish. The music was exciting and vibrant. A Flamenco dancer crashed the castanets and danced with pride and gusto. In traditional dress and with her black, shiny hair tied flat to her head in a bun, she brought the audience to a frenzy with her swirling dress and long, bare legs.

Then, as abruptly as she had begun, in a final burst of guitar and castanets she threw her head back and stopped dead on the floor.

I jumped to my feet and applauded.

My gesture was appreciated, as she took a bow and blew me a kiss across the palm of her hand.

"Gracias, senor."

After loving every minute of it, we left the bar early in the morning with a feeling that we had seen a piece of real Spanish culture.

I told Diane as we walked home through the dark deserted streets, "That was the best night I've had since I've been here. Thanks."

"I'm glad you enjoyed it," she said. "We can do it again, if you like." "For sure," I said. "I think Brian and Natalie had a good time, too."

"Actually, Natalie asked me if she could have a little time alone with Brian. She really likes him. She said she was going to have coffee with him in the apartment. Would it be OK if we had a drink at your place to give them a little privacy?" she asked.

"Of course," I told her. "I'd like that, too."

We reached the lift and, as it began to rise up, we kissed. It stopped at my floor and the two of us got out. Brian and Natalie continued up.

• • • • • • • • •

Brian was his old cheerful self the next day. We'd both told the girls that we would do it again but made no plans. I didn't want to get tied down. After a day on the beach, we grabbed a couple of spit-roast chickens and dined on the balcony with our cava.

"Let's get out early," I said to Brian. "Just in case the girls call." "Well, I don't mind if they do," he told me honestly.

"No probs, mate," I said. "I'll be at *Uncle Willie's*. If they don't call, you can catch me there or Vince's. Just don't bring them to *Uncle Willie's* if they do. Go somewhere else."

He seemed put out.

"Like I said before," I told him. "Just do your own thing. I'll see you later."

Uncle Willie's was quiet. It was still early. A few people were sitting on bar stools. Most seemed to be on their own. Some were familiar, some not. I like meeting different people in bars. You can sometimes have some great, interesting conversations with a perfect stranger.

Where you first sit can turn out to be crucial to the evening. You can meet someone and click, or you can get stuck with a terrible bore. My eyes were drawn immediately to the girl at the end talking to Uncle Willie.

She was young, small, slim and darkly tanned. She had shoulder-length, black hair and wore a vest that had no sides in it. She had no bra and, as she leaned on the bar nursing her drink, there was a tantalising show of firm, pert breasts. I can't tell you why it is, but you can look at topless girls all day on the beach and think nothing of it, unless they are spectacularly different. However, one nice pair in a side-less vest becomes suddenly exciting.

I greeted Uncle Willie with exaggerated zeal. "Buenas noches."

"Hola, Des," he cried, thrusting forward with his hand. "It is good to see you, my friend." "Likewise," said I, shaking his hand. "Is this stool taken?"

"No, no," said the girl, in a Dublin accent, "Help yourself."

"Thank you. Cevesa La Grande," I ordered. "Will you have one, Willie?" "A small beer, thank you Des," he said, pouring my pint.

To the girl I said, "what's yours, senorita? Gin and tonic?" "To be sure," she laughed. "I'm Rosie."

"How's the crack?" I asked.

"You know a bit of the brogue." She said.

"My whole family is Irish." I told her.

"You sound more like a Brummie to me," she said, leaning back to appraise me.

"Spent most of my life there. Never been to Ireland. Mom and Dad left before I was born."

"A plastic Paddy." Rosie said.

"Not even that." I replied.

She went on to tell me that she had come away for two weeks with her boyfriend, and they had had a massive row and split up. There was a party of them and it was really awkward. They were in apartments, and she had spent the last couple of days grabbing a couch wherever she could. I told her that I had moved over here for a bit, after travelling in America, and had a friend of mine staying with me for a couple of weeks. I told her how I had done quite well salvaging sunken shipwrecks and was intending to head to the Caribbean in the winter to do some salvage on a Spanish galleon that I had researched while in Barbados a couple of years ago.

Open-mouthed, she listened to my tales of daring and my close shave with three hungry sharks while diving in the Blue Grotto in Malta. For beauty, though, I confessed, I thought the Red Sea was by far the most wonderful place to dive.

"How exciting," she sighed. "What a great life."

"It has its moments."

We were laughing the night away when Brian came in. He was looking for me, I could see.

"Brian," I called.

He saw me and came over.

"This is Brian," I told Rosie. "He's the one who's staying with me. Brian, this is Rosie. I've been boring her to death all night. I'm sure she'll be relieved if you join us. Beer?"

The stools were all taken, so Brian stood with us. He blended comfortably in and we were having a great time. Suddenly Rosie gasped.

"Shit, it's my boyfriend."

Looking across the room, I could see four young men entering the bar.

They looked happy and were laughing and joking. They jostled their way to the bar and ordered drinks. They seemed to be scouring the place for a table. No chance. It was standing room only. The balcony was full and young people were sitting on either side of the steps down to the street. They saw Rosie and wove a passage through the crowd towards her.

"I don't want any trouble, Jack," she said, in a frightened voice.

"Why should there be any trouble, you old slag?" Jack spat. "You're sod-all to do with me any more. Who's your friend?"

"Des," I answered for her, "But it would be a bit of an exaggeration to call us friends. We've just met and were having a civilised chat. That is until you came along, of course."

"It's a smartarse you are, then," Jack growled, in his thick Irish accent. "Maybe I should wipe that smug little grin from yer face, you smug little bastard."

"I suppose you could try," I bluffed, watching Brian tensing for action. This was his thing. You don't do the doors of Birmingham nightclubs unless you know what you're doing and enjoy doing it. I felt brave.

"You're a mouthy little gobshite," Jack threatened.

I sat calmly on my stool and had a drink of my beer. I wanted to frustrate him and rattle him a bit.

"Look, mate, what's your problem?" I asked him. "You just said that she's nothing to you any more. So why do you want to get beaten up and spend your holiday in hospital? I don't understand. Explain yourself."

His friends were trying to calm him down and reason with him, but he was having none of it. It's true what the Old Man always said,

"When the beer's in, the wits out."

It was then that Brian took control.

"Look, pal," he said, "We don't want any trouble in here. It's a nice, quiet bar, where people are just enjoying themselves in peace. If you have a great problem, then let's take it outside into the street. I'm sure it won't take long to clarify the situation."

I smiled at that. I knew Brian to have been a very handy boxer in his early days and was a well-respected bouncer. However, we both hoped that it would not come to a brawl. Especially me. We were beginning to get attention from behind the bar and the people in the immediate vicinity.

There was no shutting Jack up.

"OK, pal," said Brian, "Let's quietly leave."

Brian made room for me and Rosie to go first. He whispered in my ear, "Meet me at Vince's. We'll split up."

Jack was behind Brian, with his three friends reluctantly in the rear. At the top of the steps to the street, Brian stopped until I was on the sidewalk. Suddenly, without warning, Brian reached back over his head and grabbed Jack by his collar and, arching down forward, he brought Jack over his back down the steps, scattering people and shattering glasses as he tumbled down. Before anyone could react, Brian was down the steps and running.

"Get the hell away," he shouted to me, as he ran off in the opposite direction. Not needing to be told twice, I was off, dragging Rosie with me.

Without further incident we made it to *Vince's bar* and were sitting having a laugh and a drink when Brian, rather breathlessly, plonked himself on the stool next to me.

"You're a prat, you are, Des," he gasped. "You start all these things then piss off and leave me to sort it out."

"You told me to leg it," I protested.

"Yes, but you wound him up."

"I never"

"Yes, you did."

"Oh, shut up. Have a drink."

Of course, this left me with another problem. What to do with Rosie. I discussed it quietly with her and formulated a plan.

"Brian," I said. "You must be exhausted. You chill out here for a couple of hours and I'll see Rosie home."

"You be all right on your own?" "Sure. No problem."

Of course, it was only after I left the bar I realised that Rosie had nowhere to stay that night.

"You better stay at mine tonight," I told her. "Won't Brian mind?" she said, a little worried.

"It's nothing to do with him," I lied. "It's my flat. Don't worry about it. He won't be back for hours. You'll be asleep by then."

There was no couch in the apartment, only two single beds, so I asked Rosie if she minded sharing. She said that was fine and was very grateful. I said I thought it best to get in early before Brian got back.

It turned out to be quite a tight squeeze.

• • • • • • • • • •

Brian didn't seem too surprised to find Rosie in bed when he got home. We cooked some breakfast in the morning and offered to walk with her halfway to the beach.

Also on their way to the beach were Diane and Natalie.

"Morning, Des," Diane said, a note of sarcasm in her voice as she eyed Rosie in her evening clothes. "Have a good night, did we?"

"Actually, no," I countered. "We got into a bit of trouble with Brian's friend, Rosie. She had a slight domestic problem and had nowhere to stay, so she had to doss with us for the night."

"Oh, I see," said Natalie, giving Brian daggers. "Well, have a nice day."

With that they were gone. So were we. We took Rosie as far as we were going and wished her the best of luck.

I hoped that was the last we were going to see of her.

The days were running out for Brian so, after our dinner and cava with strawberries and ice cream, he announced,

"I think I'll pop up and see the girls. Ask them if they fancy a night out."

"Good luck with that one," I smiled. "I think they might have got the hump over Rosie." "She was nothing to do with me," he protested. "That's you again. You always manage to duck out of everything."

"Just tell them the truth. Her boyfriend was going to kill her in *Uncle Willie's* and you came to her rescue. She was too scared to go back, so we let her have my bed and I slept on some cushions on the floor. Anyway, that's exactly what happened, right?!"

"More or less," he conceded. "The only thing I don't remember is you sleeping on the floor."

"A minor detail. Don't worry about it." "I'll pop up now. See you in a bit." "Okey-doke."

He wasn't gone long.

"They've already made plans."

"Really?" I was surprised. "Just as well, really. We need to sort things out with Uncle Willie. We never paid the bill. At least he knows it wasn't our fault. They came in looking for trouble. In fact, it could have been a whole lot worse for him. If we hadn't taken them outside, they could have wrecked the bar."

"I suppose by the end of your spin on it, he'll owe us and pay for the drinks," Brian scoffed.

"Never thought of it that way," I said. "Of course. What would have happened if we weren't there?"

"Sounds about right."

"Anyway, I think I should call in the *Principe Sol* and see Jane," I told him thoughtfully. "I've missed her."

"I'm sure."

"Tell you what; I'll get over there early and see you later at *Uncle Willie's*."

There are no regular hours for a holiday rep and, fortunately, Jane was finished early that evening. There were no new arrivals to take up time asking mindlessly stupid questions. She would see me in *Uncle Willie's* for a few drinks. I cleared things up with Willie. I paid our bill for the previous night and all was well. Jane wanted an early night.

"Can I see you home, then?" I asked.

"Of course, Des, why not?"

"I wouldn't want you to think I was taking advantage," I said sincerely. She pressed her finger to my lips.

"Shush, Des. Say no more. We both know the score. What are friends for?"

• • • • • • • • •

It had finally come, Brian's last day. We were going out with Diane and Natalie. They suggested the Spanish bar and we were all for it. Not only was it a great bar, but it was also really cheap compared to some of the tourist bars.

The night was much the same as the previous one, with Brian ending up staying upstairs with Natalie. I'd spoken to Jane about Brian's flight, and she arranged for him to get a lift on one of the coaches to the airport. After a few drinks in the *Principe Sol*, I saw Brian off.

Thirty-seven

It was a relief to see him go. I didn't like the restriction of having to worry about pleasing someone else. I was now free to do what I really wanted to do. Blend in with the locals and the workers. It was like a little community. They all looked out for each other. They came from all over the country, Scotland, Liverpool, Newcastle, Birmingham, Dublin. I was the only one from Jersey. They all worked somewhere, a hotel or a bar, or sold tickets for excursions or nightclubs.

I was never lonely. There was always a friend to be found somewhere. One night, in a video-cum-music bar somewhere in town, *Top of the Pops* was on. You can imagine how excited I was when some friends of mine from Small Heath in Birmingham had made it to No 1 in the singles charts.

UB40 with *Red, Red, Wine.* Raymond and Earl Faulkner had been friends of mine for years. Their mom and dad had been customers of mine for as long as I can remember. Raymond's love of my beef burgers was legendary. Once, on returning from Los Angeles, the first thing he did was walk down the hill from his mom's house to the shop and buy four quarter-pounders. After a catch-up chat about his adventures with the band he tells me,

"I'm sorry, Des, I haven't got anything smaller than a fifty."

"Don't be a prat," I told him. "I can't change a fifty-pound note for a few burgers. Owe me!"

Tragically, not long after, he died in a car crash, the car being driven by his brother Earl.

My street cred went through the roof, as I made no secret of the fact that they were mates of mine. I must confess, also, that I may have over-egged it just a bit.

During my second week, I met a couple of really nice lads from London. They worked in a bank and were on holiday for two weeks after collecting large bonuses from the city. They were staying in a hotel on the front, not far from the *Principe Sol*.

I was seeing a lot of Pete and Dave when, one night, they told me about an excursion they were going on. It was to a place out in the hills called Malaga Lakes. It sounded exciting and different. I asked if I could go with them. They said they would speak to their rep and see if there was any room left on the coach.

There was. I gave them some money and they sorted it. It was going to be an early start from their hotel. The night before, we had a good drink in *Uncle Willie's* and it was agreed that I should bring a change of clothes for the morning and doss in their room. I slept on the floor with some cushions and a blanket.

In the morning, I showered and changed into T-shirt and shorts and went with Pete and Dave down to breakfast. It was a huge, help-yourself buffet of hot and cold food. There was no one checking who went in and out, so I had a good feed to last me for the day.

We boarded the coach about nine. It was a fairly long ride through rural Spain. We passed many small villages and farmhouses until the coach swept into a large, rock-and-gravel car-park. There were already a few coaches there, parked in a line. We were overlooking a huge, vast lake.

On it already were a few long, canoe-type boats with oars, and about eight people in each. The name of the game seemed to be to soak as many people as you could in the nearest boat. By splashing your oars and throwing water with anything you had, the people in the next boat were howling with laughter and soaking you back.

Up a hill, on a plateau, was a large bar restaurant. It was approaching noon, so we walked up the ramp that led to it and found a table overlooking the fun. It was another hot, Spanish day with not a cloud in the sky.

For a long time, I had been working in my mind on the plot for the first book I was going to write when I retired. The plot involved a small private jet, full of stolen gold, crashing into a lake somewhere and being discovered years later, untouched since the day it crashed.

I knew at that moment I had found the perfect location. There was no fishing or diving permitted here and, apart from these holiday excursions, a plane sunk here could remain hidden and undiscovered for years.

The book was to be called *"Private Execution"* and the gold was to be from the famous Brinks Mat robbery. Although the book was years from fruition, Malaga Lakes left a huge impression on me. It was, in a way, going to help to change my life.

I declined the chance to get soaked and, instead, ordered a bottle of cava and a bowl of strawberries and ice cream, while Pete and Dave had fun on the lake. I was happy in my own world.

At the end of the third week of my breakdown, Diane had to fly home. Natalie was a little lost. She wasn't very good on her own. I told her she could pop in any time she liked for a drink. After all, it became a little tedious going out every night and, sometimes, it was nice just to stay in with some food and a drink.

A couple of times I visited Jane and met up with her at *Uncle Willie's* with Dave and Pete. One time, Natalie came and we all five had a great night. I walked Jane to her place with Natalie and explained that, as she lived so far away upstairs from me, I felt duty-bound to see her home safely.

Jane said she understood perfectly.

"After all, Des," she said, "What are friends for?"

Rather awkwardly, I saw her to the lift and kissed her goodnight. "See you soon, Jane."

"Sure, Des, see you soon. Say no more, we both know the score. Just don't forget!"

"I won't."

I didn't, of course, but the last week fled. My last afternoon I was swimming and sunbathing at Pete and Dave's hotel. I decided to take a running dive into the pool when I slipped on the wet concrete and, instead of landing in the water, I landed on the floor. I was pretty badly hurt and bleeding heavily from my knees and arms. I had to be carried into the hotel, where I was given first aid.

They did a pretty good job on me, I must say. They cleaned me all up and, using a few butterfly stitches and smothering thick, yellow stuff all over the cuts, they bandaged me up pretty good. It was hard to walk. They called me a taxi and took me back to the apartment. I limped to the lift and got back to the door. It would be a night in tonight.

I phoned Natalie upstairs. She came straight down. With no plans for the evening, she offered to cook me some food and look after me. She stayed with me all night, God bless her, and helped me to pack my case in the morning. I'd already decided I was getting a taxi to the airport, as I didn't want to piss about with Jane and coaches.

There was virtually no food left, and I told Natalie to take whatever drink there was upstairs.

I'd already said goodbye to Pete and Dave and exchanged details. They were good friends and were later to come and stay with me and Rita at the house.

I also exchanged details with Natalie. She was also to come and stay with me for a weekend. Rather than hang around here, I got Natalie to call a taxi that they always used. They were far cheaper, she told me. I left the key for Vince with her as arranged, and she saw me off to the airport. I was hours too early but would rather be there and settled.

Although quite stiff, my legs weren't bad at all. That is, until I reached the check-in desk. The flight wasn't open yet but, being very sympathetic to my plight, the two Spanish senoritas offered to keep my case behind their desk. Seeing that I was in need of plenty of legroom, they allocated me a seat in row one at the front by the door. Also, they said it would be easier for the aircrew to take care of me during the flight.

I thanked them for their kindness and limped off to the bar. The flight was not full and I found myself the only person in the front row. I wasn't always alone, however, as, quite often, one of the stewardesses would sit with me and ask if I was all right and was there anything I needed?

"Well there is, actually," I said, "But I suppose that's out of the question?" "Bad boy," she laughed.

Thirty-eight

Back home again. It was great to see Rita and Angela. They were glad to see me and I had lots to tell them, and lots to not. I came back with £250 in American Express travellers' cheques. I had lived well for four weeks on £500. I decided not to cash them in. You lost a lot in the exchange and only ended up paying more money to buy them back the next time you went away.

Rita had collected the money owed from both shops and been OK. She also had a substantial amount over for me. I reflected that, with all the trouble I had to come, I would have been more than happy to have lived on in Spain. I had lived well without work, so imagine how well you could live with a part-time job. There were at least half a dozen bars where I was guaranteed work. I'd got the gab for it, I'd been told. Well, really, it's not much different from selling meat. There were also many English bar-owners who weren't happy with the quality of the meat they were getting. They would be happy to pay someone to oversee the deliveries, or even go direct to the market. No problem to me.

Instead, I was facing a cold, English winter, another Christmas of pulling my hair out over turkeys, and a New Year court appearance that was almost certainly going to end with a three-year driving ban.

Stuff that. If only. I'd only been back a few weeks and already I was missing my carefree life in Torremolinos.

I got through the New Year. The priority now was my trial.

I met my solicitor, Howard Joy, at the *New Inns* in Small Heath, where I often met Ray Faulkner from UB40. The band did a lot for the youth centre across the street and used the pub for a beer afterwards.

Howard asked me my thoughts on the case. If I pleaded guilty before the day, I could save myself a lot of money. The delays had served their purpose. The mandatory sentence for a second offence was a three-year ban. So, unless I got off, which was highly unlikely, what was the point in fighting it?

He slapped my file on the table with a couple of pints.

"So, what do you think, Des?" he said, still in his court clothes. "If you fight it, you'll need a barrister. I know one that will do it for cash in the hand. He's good, but he still won't be cheap, even at half price."

I knew he was right. However, I was always pissed off by the unfair way I was caught and wanted my say in court. I wanted to go down fighting. I took the folder and started to read the file.

Most of it I already knew. There were statements from the arresting officer, the desk sergeant who took the second breathalyser and a letter from the doctor who gave me the blood test, with the results.

I noticed something wrong. The blood sample.

Results of blood taken from Mr Desmond McGRAPH on Dec 26th.

"Who's this bloke?" I asked Howard Joy LLB. "I don't know him. He's not me. Why have I got his blood results?"

"What do you mean?" asked my pound-a-minute LLB solicitor.

"This geezer McGraph, what's he got to do with me? They've obviously mixed me up with somebody else."

Howard Joy LLB, Senior Partner at Southall & Co Solicitors, stared blankly at the results. "What do you mean?"

"McGraph," I told him, pushing the evidence in front of him. "I'm McGrath, not McGraph. Who's he?"

"Bloody hell, Des," he gasped, "You might be on to something there."

"I thought that was your job," I told him, finishing my beer. "Mine's a Brew XI."

It was a slim hope but, even if I could make someone look stupid in court, I wanted to take the chance. I might go down fighting, after all.

"Get me the barrister," was my war cry. "Let's give them hell."

A meeting was arranged at the *Pen and Wig*, a pub by the Law Courts. It was late afternoon. I knew Howard was a piss-head, so it followed that the barrister would be, too. I wasn't wrong. I'm not going to write his real name. Not because of legal reasons, you understand. Just that I haven't got a clue what it was. All I can remember is that he was small, with a red face full of broken veins, and was a pig farmer as well as a barrister. When he found out I was a butcher, he seemed more interested in selling me a pig than getting me off.

I pointed out that I hoped we were not on the clock with the pig conversation. "Of course not, old boy," he spluttered. "Now, where were we again?"

To be fair, it was a hopeful and entertaining afternoon. After studying all the evidence, he also had seen something else that could possibly weaken the prosecution case. He would take it all under review and post me what he called a full "Barrister's Opinion".

I would wait with bated breath for *Rumpole of the Bailey's* expert opinion.

We can give them a fight, I was told. We have one major surprise for them and another one they won't have seen coming. He went on to explain. The law stated that a breathalyser had to be taken in or near the car.

In 19 something or other, in the case of the Crown versus Fred Blogs, it was decided that 150 yards was not in or near the vehicle, and the case was dismissed. He had noted that when I had taken a walk, with the arresting officer's permission (to take a leak), was I more than 150 yards from the vehicle? I said I thought so. He told me to find out and, if so, get a professional-looking drawing to scale proving it.

I did just that. I was over 200 yards from the vehicle when breathalysed. My drawing showed where the car was on the road, the surrounding area, and a cross marked the spot where I was breathalysed.

We were ready for court.

It was a dark, bleak, miserable February afternoon. Perfect for an execution. I met Rumpole in the courts and we finalised our battle plan. He asked me about the copper who arrested me. Did he seem bright or dumb?

"About as bright as a turd in a bog," I answered honestly.

"That's good," he said, "Because I'm going to ask him a simple, straightforward question.

If he gives the right answer, he's trapped and can't go back on it." I asked him what it was.

"Just watch and enjoy," he smiled.

I'd told Howard that I couldn't afford two legal brains and we could manage without him. It was just me and Rumpole.

It was time. I took a breath, Rumpole adjusted his wig and squirted some freshener into his mouth, and we all stood up for the magistrates.

I pleaded not guilty.

The prosecution outlined their watertight case, concluding that I was wasting the court's valuable time and deserved nothing less than beheading in the Tower of London.

"Well, if that's the case, your worships, what's the point in my client having a defence lawyer? Why don't you just lock him up and throw away the key?"

"I protest," howled the prosecution.

I had to smile. I'm sure it's all a load of bullshit. A game they play among themselves. You see them afterwards in the pub, laughing and joking together while they knock back their gin and tonics and whiskeys.

Rumpole called the arresting officer and, with a rolled-up drawing of the crime scene under his arm, he began questioning. He entered the drawing into evidence and showed it to the witness.

"Would you agree, Constable, that this is an accurate drawing of the area where you stopped Mr McGrath?" he asked.

Dixon of Dock Green studied it carefully and agreed that it was.

"Would you agree that the map shows accurately where Mr McGrath's car was and where he was when the Panda car arrived to breathalyse him?"

Dixon took his time and eventually agreed that, yes, all was in order.

"Thank you, officer," said Rumpole. Then, addressing the magistrates, he pointed out that I was, indeed, 225 yards from the vehicle. He then produced the law book that confirmed the case of Fred Blogs. He concluded that I was too far from the vehicle and, in fact, the breathalyser was illegal.

The prosecution looked shocked and blustered on a bit. Round one to Rumpole.

There was a bit more jousting between counsels, when Rumpole stood rigid with a look of horror and shock on his face. Holding up a piece of evidence, he gasped,

"Your Worships. Whose blood sample is this? It's not my client's. It belongs to someone called McGraph. Look," he said, handing it to them.

The prosecution were on their feet. They examined the evidence. "It's obviously a spelling error," they protested.

"I'm sorry, but you can't convict my client on somebody else's blood results." Chaos ensued for a few minutes.

The magistrates agreed with Rumpole. The prosecution wanted time to call the original doctor as a witness to identify me. They wanted an adjournment. The magistrates said no way. They'd had over six months to prepare their case. They could produce the doctor this afternoon. If not, well?

Round two to Rumpole.

There was panic in the air. We couldn't imagine how they were going to do it, but they were pulling out all the stops to get him and bring him to court.

The case was stopped for a bit and we had to hang around for two hours, biting our nails and just hoping they couldn't produce the doctor.

With only minutes to spare, they dragged the bastard in. He looked like a kidnap victim. The trial resumed.

I'd had my moustache since I was a teenager. I'd never been without it. I couldn't imagine what I would look like without it. I didn't look good! I'd taken advantage of my time and found a razor and shaved it off.

I didn't recognise myself!

The doctor stood in the witness box and looked at me.

"Can you identify the defendant as the person you took a blood sample from on December

26th last year at Stechford Police Station?" the prosecutor asked hopefully.

The doctor studied me carefully.

"He looks familiar," he said, thoughtfully. "But I couldn't possible say he was the man I took a sample from. It was over a year ago."

He was being honest and truthful. As we had hoped. How could he have positively identified me? Rumpole did his job and made his point that the whole exercise had been a complete waste of time. They couldn't prove the blood sample was mine. End of story. No case to answer.

The magistrates were mumbling in hushed tones and said they were retiring to consider their verdict.

They were out for ages.

Rumpole had a nip from his small hip-flask and smoked one of his small cigars. He thought the longer they were out, the better it was for me.

I was getting really hopeful.

The Law Courts were deserted.

It was late and gone closing. They finally came back. They asked me to stand.

They told me that they had thought long and hard before coming to their decision.

Unfortunately, they found me guilty.

Without calling for the black hat, they said that, reluctant as they were, the only sentence they were allowed to pass by law was a three-year driving ban. They fined me the minimum they could, and actually said how sorry they were.

I politely thanked them.

Game, set and match to the prosecution.

I was pissed off.

It would take weeks to grow back my moustache. Rumpole thought I had grounds to appeal.

I told him to forget it.

It was over.

It was time to draw a line under it all and get on with it. Spilt milk and all that.

I took a taxi to *The Fox* in Walmley and let it all sink in. Three years was a long time to mope about it. I'd overcome worse things in my life and I could certainly overcome this. It wasn't as if I was in prison. I didn't need a car to get to the pub. I needed a car to get to work and run my business.

I'd had a year to think about it. Now was the time to do it.

Thirty-nine

I thought about all the positives. I would save a fortune not running the Jensen. There was no sense in keeping it for three years to rot in the garage. Jensen had gone bust and were no longer in production. Good examples were going up in value. I'd paid five grand for it and sold it to a dealer for eight and a half.

Nice profit.

The boat had always been a white elephant but was virtually brand-new. I sold that for two and a half, breaking even. Rita had the MG, which was paid for, so we still had a nice car. The van owed me nothing and would see me through with Rita driving it.

The market traders were all great mates and I knew, from the last time, they would help me as much as they could with deliveries. It wasn't going to be easy, but there was life after the breathalyser.

Pat Cody was doing well at the Broadway shop and was there every week with my rent. He had a network of friends in the wholesale business and was always getting great deals bulk-buying. He couldn't shift it all himself and, between us, we shared some good money.

Trevor and Les were beavering away in Somerville Road. I didn't tell Trevor, of course, but I planned to get through the first few months and take back control of the shop myself. It would mean making him redundant. When the time did come, he was all right about it and had always planned to get a shop of his own anyway. This he did and went on to have a very successful business of his own. I was glad for him. We always remained good friends and, whenever we met down the market, he always called me Mr Des.

So, with this new, positive attitude, I did what I'm best at. I booked a holiday. I don't know where we went. I can't remember. But it set the theme for the next three years. I went on holiday three times a year. Twice with Rita and once on my own.

I always set aside Easter for myself and a mate.

My brother, Ed, was great and came to my rescue whenever I needed him. We became even closer, if that were possible. We played golf and squash and found ourselves frequenting a lot of wine bars. They were the in thing of the day. Somehow, our conversations always drifted back to our road trip in America.

One afternoon, after he had taken me home, I was sitting in the back garden reading the *Daily Mail* when a particular advert caught my eye. It was a piece in the property section. A company in Bradford was promoting properties in Florida. Apparently, the favourable exchange rate between the dollar and the pound meant that you could buy a home in America for a fraction of what it would cost normally.

The company was Stephen Hardy and Partners. It had a development of two-bed bungalows with pool that it was selling for £16,000. It sounded amazing. I phoned up and received a package in the post with all the details.

The Sirocco was a detached, two-bed, two-bath property with swimming pool. It was located on a waterfront development near Clearwater, by the Gulf of Mexico. The pictures looked enticing. I phoned up and spoke to Stephen Hardy himself. I told him how interested I was and he explained that they were doing five-day free inspection flights. He would send me further details.

They arrived and sounded too good to be true. The package included free hotel accommodation at the Ramada Inn, Countryside. I was, of course, immediately interested.

I was devising a plan. Talking to Ed about it, like myself he was sceptical but would be interested. I spoke to Stephen Hardy and explained to him that my brother and I had butchers' shops in Birmingham and properties that we rented out. I told him how I was intending to take a year off to write a book that my publishers had given me an advance on. He was quite fascinated. I told him that my original plan had been to rent a house in Florida for a year. However, after considering the purchase of a Sirocco, it made far more economic sense to buy one of those and sell it on after a year, or even keep it as an investment.

"You'd be foolish to rent, Des," he told me over the phone. "The projected resale value of the Sirocco, after a year, would make it crazy to waste money renting."

"That's exactly what I thought," I said. "Also, I was telling my brother about it and he said he'd love to buy one, too. He wouldn't be living in it, of course, but would be wanting to rent it out. I said that if he did that,

I'd be there to look after it for him."

"Sounds like a great idea, Des," he said, enthusiastically. "There's only one thing that bothers me, though," I said. "What's that?" he asked.

"Well, realistically," I continued, "It doesn't seem practical to go over for five days. It's a long way. You can't do these things overnight. It wouldn't be worth going for less than ten days."

"That's no problem," he said, dismissing my concern. "We can easily arrange a ten-day inspection. Don't worry about that."

"Well, that sounds great," I said. "Let me talk to my brother and I'll get back to you."

I discussed it at length with Ed and he said it all sounded great and he would leave all the arrangements to me. He still had serious reservations but, if I was happy with it all, he was up for it.

To be honest, I thought there had to be a catch somewhere. I spoke at length to Stephen Hardy and eventually set a date in October. We were to fly direct to Tampa from Heathrow with Pan Am. We would be met at the airport and taken to our hotel in Countryside. At no time was there any mention of paying any money.

I reported back to Ed.

"Still don't get it," he said. "Me neither," I shrugged.

The tickets arrived in the post, along with luggage labels and a brochure of the Ramada Inn.

It looked fabulous. Too good to be true!

"Well, Des," said Ed, "I can't see what we've got to lose."

Ed had to break the news to the Old Man, who wasn't very happy. I made all the arrangements I needed to with the shops, and the time came around.

With plenty of travellers' cheques and my trusty American Express card, we took the coach to Heathrow. There was a great air of excitement about the whole trip. We still thought there must be a sting in the tale. Dismissing any negative vibes, we checked in at the Pan Am desk and found the bar. After a few beers, our flight was called and we boarded where, to our delight, we discovered it was a free bar.

• • • • • • • • • •

It was a great flight. We had no car to worry about, so we were able to take full advantage of the refreshments. The meals were great and, after landing, we had no problems with Immigration. I think that, after our previous visit to the States, they must have had most of our details.

Our instructions were to pass through Arrivals and look for someone holding up a card with our names on. True to their word, as we walked towards the final glass wall of doors to the outside, we saw a young man, mid-twenties, holding up a sign - McGrath.

We approached him and introduced ourselves. Holding out his hand, he smiled broadly. "Nice to meet you guys. My name's Ed, Ed Purser. I'm your driver. How was your flight?"

"Couldn't be better," I told him truthfully. "I'm Des. This is my brother, Ed." "Two Eds," he joked, "Better than one."

I liked Ed Purser. He was tall, about six feet, and athletic. He wore long pants and a white, short-sleeve shirt. He took us a short walk to the car. It was a large, black Lincoln Town Car.

Impressive. He loaded both our cases and hand luggage into the boot. Before he could close it, we tossed in our jackets. It was mid-afternoon, and hot.

We climbed in the back of the car and were relieved to feel the cool air-con.

"It's about a half-hour drive to the Ramada," announced Ed. "I'll give you the commentary if you like, or just point out anything of interest if you prefer."

"The interest bit will suit me fine, Ed," I said. "What can you tell us about where we're staying?"

"Sure thing," he began. "The Ramada Inn is a really cool place. The rooms are large and high-grade. It has a really well-respected restaurant and a great bar, with live music most nights. *Jack's Place*, it's called. You get a lot of hot chicks hanging out there at night. Countryside is a small town with a huge, new, shopping mall. There are loads of bars, restaurants and shops for you to spend your dollars in."

"Sounds great."

"It is, but I don't think you guys will have all that much time on your hands. The real estate tours are pretty intense. The sales people are on the case pretty much twenty-four seven.

Ed and I exchanged glances.

The road turned from land to sea. Ed Purser told us we were driving over one of the largest causeways in America. Ships and boats of all sizes passed below us. We soon drove into the front gardens of the Ramada Inn. Palm trees, water features and fountains. Most impressive.

The car stopped outside the huge glass doors. Ed popped the boot and we collected our baggage. He showed us to Reception, spoke to the girl on the desk and told us he would leave us in peace to settle in.

"I'll see you guys later," was his farewell. "Thanks," we both told him.

A smart young man in uniform loaded our cases on a cart and we followed him to our ground-floor room. We opened the door and he followed us in, stacking our baggage by the wall.

I knew the drill and gave him a tip. The room was huge, with two double beds. A wide, full-length balcony overlooked a huge tropical garden area. In the distance was a very large pool and jacuzzi.

"Bloody hell," gasped Ed, "This is some place. It must cost a fortune."

"Well, someone has to pay for it," I said. "There's got to be a real hard sell somewhere along the way. We'll have to keep sharp and have our wits about us."

"What about we find the bar and get our bearings?" suggested Ed.

"Good idea, Ed," I said. "Have you noticed a theme developing here again?" "What's that, Des?"

"We always seem to end up making plans in a bar." "Great source of inspiration, Des."

"Roger that."

I can't really remember any moment of great inspiration. There were very few early-evening drinkers. *Jack's Place* was a fabulously laid-out tropical garden theme. It had the James Bond, Caribbean and pirate treasure feel all in one. We loved it and could see a good few nights passing before us, cocktails in hand.

Tiredness was beginning to catch up on us and, after one for the road, we went back to our room. An invitation was under the door. Could we meet in the bar at 8.00pm for cocktails before dinner?

• • • • • • • • • •

After a short sleep, we unpacked our clothes and had a shower. I pulled a couple of Buds from the fridge and we drank them by the neck. Selecting smart but casual clothes, we went to the bar. It was busier than before.

Almost instantly we were approached by a short man in a suit and tie, who surprised me when he spoke, being English. He had no accent. His hand shot out and I took it.

I'm Bill, Stephen Hardy's partner. How are you both?" "Very well, thanks. I'm Des, this is my brother, Ed."

Ed held out his hand. Bill swapped. "Nice to meet you," Ed said.

"Stephen's told me all about you," Bill enthused. "What can I get you?"

We told him beers. He went on to tell us that there were eight clients in all on this particular inspection trip. His American partner, Charlie, would meet us with the others in the restaurant. As it was our first night, all would be kept casual and low-key, letting us get the feel of things.

Fine, we said.

Bill ushered us in to the restaurant where, at the far end, was a table for ten. The other six clients were on either side of a long table. At the far end, seated like the chairman of the board, was a middle-aged man in a suit and tie that was obviously not big enough for him. He was overweight and looked uncomfortable.

As we approached, he got up to meet us. He said jovially, "Hi, I'm Charlie; you must be Des and Eddy. Glad to have you guys on board. Settle yourselves down."

We sat facing each other at the end of the table. Bill had his seat by us on the end facing Charlie. The other six clients were all English, and retired, I would have said. It took a few minutes, but we all sat swapping names that I instantly couldn't remember.

Charlie told us all to select our food. The waitress came and took our orders and each couple selected a bottle of wine of their choice. Charlie gave us all the spiel about his company and the amazing opportunities available to us. He predicted a massive growth in the housing market and pointed out tremendous opportunities for the lucky people who managed to get in at ground level.

With values soaring rapidly, he couldn't emphasise enough to spend big now and profit big later. He was very persuasive and it was hard to argue with his logic. I told Bill so.

"I'm impressed, Bill. I'm glad we came already. I can't wait to see the Sirocco. If it's as good as you say, I'm sure you'll have two very satisfied customers."

"Sixteen thousand pounds," our Ed joined in. "At the present rate of exchange, that's about thirty thousand dollars. Yes?"

"Pretty much bang on," agreed Bill. "Are there any extras?" asked Ed. "Just sales tax, a couple of percent."

"So, the whole shooting-match will be south of twenty thousand quid?" "Easily," agreed Bill.

After the meal we all went our separate ways. Us to the bar.

We had come away with no intention whatsoever of actually buying a place. We just wanted the free trip. However, we had to admit we were getting drawn in. We looked at it all over again at the bar.

The Sirocco was a detached, two-bed, two-bath with a pool. At under ten grand each, it seemed a steal. Between us we could easily afford it.

We were starting to get excited.

The next morning after breakfast, there was a small bus waiting for us, to show us the sights and developments available. Ed Purser was the driver and Charlie was doing the commentary.

Florida, I can safely say, is the most beautiful place to be in the world. If you love clear blue skies, golden beaches, marinas, boats and waterways littered with wonderful bars and restaurants, then Florida is for you. Most of the morning and afternoon we spent wandering around fabulous locations with unbelievable homes at prices a fraction of what they cost at home.

There were detached homes and gated communities of low-rise condominiums. Flats to us. However, there was nothing anywhere near our price range. Not that we were bothered, we were enjoying the sights. We stopped by a waterfront bar and Charlie told us to take a break for an hour. We did.

That night, it had been arranged for us all to go to a Greek restaurant.

The food was great, and all free. We listened to the talk of where we were going the next day and the property we would see.

The next day was the same as the last. The only difference was we were going to a Mexican restaurant that night. The next two days mirrored the last two, except one night we went to an Italian, then a Japanese.

The Japanese food was revolting. I took Charlie to one side and told him so. I said we were getting a bit fed up and wanted to see something in our price range. He agreed and arranged for Ed Purser to drive us to some developments that were not so close to the waterfront. Waterfront homes cost double the ones in communities more inland.

After breakfast the next day, the six other clients went in the bus with Bill and Charlie, while we went in the car with Ed Purser.

Free of everyone else, it was nice to relax and have a chat to Ed the driver. He took us to about four different condos, way out of our price range.

"Ed," I told him, "I think we're wasting each other's time here. We've seen nothing anywhere near what we can afford. We just want to see a Sirocco."

"I'm just the driver, Des," he said with a shrug. "If you guys don't want to see any more properties, why don't I find you a nice bar and drop you off there? Then I can go home and do my washing and pick you up later."

We were more than happy with that. He took us to a great marina, where there was a fantastic bar. *The Captain's Locker*. It was about one o'clock.

"What about I pick you guys up about four?" suggested Ed.

We were fine with that and spent our afternoon at the bar, with the local boat people.

Before we knew it, Ed Purser was back to take us home. He told us we were all meeting in the hotel that night and not to mention our afternoon. There was going to be a big sell afterwards in one of the small conference rooms. He told us the names of a few places to remember in case Charlie or Bill asked us where we'd been.

"No problem," we told him.

At dinner, Charlie was talking and recounting all the developments the group had been to that day. He whetted our appetite with the unmissable deal he had to show us all after dinner. There was a murmur of excitement among the clients.

Was this what it was all about, I thought?

Charlie addressed us. "And where have you guys been today?"

"Been all over the place, really," I told him. "Saw a lot of condos. I can't remember the names of the places."

I looked at Ed Purser. "Ed, can you tell Charlie where we've been? There's that many, I haven't got a clue where they were."

Charlie seemed satisfied and we all retired to a conference room, where there was a big table completely covered with a model of a luxurious-looking development.

"This is our latest project," Charlie announced with pride. "It's in the pre-construction stage."

"What does that mean?" asked our Eddy.

"It means that, even before they are started, you can purchase the best plots in the primest locations. By buying in first, when the development is completed, you will be able to sell your home at an estimated 50-100% profit. This whole development will have the finest shopping mall in the area. There are going to be three golf courses, here, here and here."

He went on to point out all the other amenities, cinemas, bowling alleys, whatever.

People were standing over the model excitedly, hanging on his every word. Reluctant as I was to piss on his parade, I asked how much it was to buy in.

I told him straight, it was way out of our price range. I asked him where, exactly, this development was. I said I'd never heard of the place. Could he point it out on a map for me? The rest of the audience were as interested as me. Charlie seemed to be getting a little agitated. The site was in Northern Florida.

"How far is that from here?" I asked. "What else is there around the place? There doesn't seem to be much waterfront."

Charlie explained everything eloquently. It was time to end the evening.

The next morning after breakfast, I had a word with Bill. I told him I was going nowhere that day. I said it was pointless showing us hundred-thousand-dollar condominiums, as we just couldn't afford one. All I wanted was to see a Sirocco. He said he would talk to Stephen.

"Fine," I said. "We'll be by the pool."

"Bill doesn't look very happy, Des," Eddy said, as we sat by the pool with a beer.

"Well, something's not right," I said. "I think the whole thing's a big con just to get you over here."

"Well, I think they've come unstuck," said Eddy.

We were standing in the pool with our backs against the side when Bill came bounding over.

"Des, Eddy, I think it's time we had a talk."

"Fine by us," we said. "We'll get changed and see you in Reception."

Bill marched off. We dried and changed into shorts and T-shirts. In Reception we waited a while. As I was checking the time, Charlie and Bill came over to us.

"Hi," said Charlie, "How are you guys today?"

"Marvellous," I beamed. "Bill said you wanted to talk to us."

"Sure," he said, "I was just wondering about your intentions."

"Well, Charlie," I began, "It's no great secret. We've come over here to buy a Sirocco.

However, all you've shown us is a complete waste of everybody's time. There's no way we can afford a luxury waterfront condo. All we want to see is a Sirocco. What's the problem, Charlie? Just show us a Sirocco."

Charlie kind of grinned. "Over to you, Bill. These guys want to see your Sirocco." The emphasis seemed to be on "your Sirocco."

Bill looked flummoxed. "Well, the problem is, the Sirocco has been discontinued." This was my opportunity.

"Are you telling me, Bill, that you've dragged us all the way over here to see a Sirocco, and you haven't even got one? I'm not very happy, Bill." I was almost shouting. "Come on, Ed, let's get out of here."

We stormed off to the bar.

"Blimey, Des," said our Ed, "You were brilliant. You should have been an actor."

"I know," I said smugly. I was still smiling when Charlie came over. He seemed amused.

"I don't think it's a good idea, you guys joining us tonight. You seem to be having a disruptive influence on the group. Why don't you guys just go in the restaurant and have dinner on us? Just sign the tab."

That was fine by us. Not wanting to seem ungrateful, we ordered the most expensive meal on the menu. Surf and Turf. Steak and lobster with a bottle of *Blue Nun*.

The rest of the evening we spent at the bar. The staff were all really friendly. Around midnight, the bar was closing. We were joined by a young, long-haired, surfer-type American kid in his mid-twenties. He was dressed casually in jeans, T-shirt and sneakers, with a short denim jacket. He introduced himself as the dish-washer. He had just finished and was going out on the town for some beers. He asked if we would like to go with him.

We thought it would be fun.

"Why not?" we told him.

"I'll get my car and see you out front," he said.

As we waited outside, a horn sounded and the dish-washer pulled up in a long, old, American car that looked in showroom condition. It was vast inside, with two full-length bench seats. The music system was disco quality. I loved the band that was playing.

"Who's the band?" I asked. "They're great."

"Foreigner, man," he told me. "You must have heard of them. They're British. Huge over here."

Eddy joined in, "I've never heard of them, either. I love them."

We were driving along a freeway for about twenty minutes when the dish-washer took the ramp into a huge retail estate with shops and a mall. We were going to a bar-cum-nightclub.

We were told that there were four stages in the club and, as one band finished, another began. The bar was located in the centre, with dance floors all around it. It was twenty dollars each to go in, but that included so many house beers free.

The atmosphere inside was fantastic. The live bands were proper American Rock. Loud guitars, pounding drums and howling singers. We sat at the central bar. The stage lights and laser beams changed from stage to stage. The air was blue with the smoke from cigarettes and weed.

The dish-washer melted away and returned with a long, thick joint glowing from his mouth. We offered him a drink and he had a beer in a bottle. He told us all he knew about each individual band as they pounded out their songs.

Ed and I were completely wrapped up in the music. We both loved heavy rock. Ed's favourite was Meat Loaf. He had his *Bat Out of Hell* album at home and drove the Old Lady mad playing it full blast in his bedroom. One of the bands hammered out some Black Sabbath. Memories came flooding back.

It was time to go, the dish-washer told us. We found the car. I sat in the front and Ed in the back. The driver was smoking another large reefer with the window down.

He was completely off his head and totally unable to drive. Eddy volunteered to take the wheel. I can't imagine what I was thinking of, but I agreed. I can only imagine it was because I was banned and had no licence and was also drunk.

Ed's effort was a total disaster. He'd never driven an automatic and was hopping and jumping all around the car park, obviously blown out of his head.

"Stop," I remember screaming. "Get out, for Christ's sake."

I got Ed in the back and the dish-washer beside me.

"If I drive, are you capable of directing us back?" I demanded.

"No problem, man," he slurred, "Just get the ramp north on the freeway to Tampa. Take the ramp off at Countryside and we're home."

After about four attempts to find the right exit from the car park to the freeway, we were following the overhead signs. Shitting myself all the way, we finally saw the big neon sign for the Ramada Inn. Parking the car where there was plenty of space, we thanked our host and left him in the car. He was going to sleep it off, he said.

• • • • • • • • • •

We had two days left and, after breakfast alone, we decided to explore the Countryside Mall that day. It was a fair old walk. In America, everywhere expects you to have a car. We could see the mall from the Ramada.

Trouble was, it was two eight-lane freeways away. Out of the hotel there were four lanes going one way and four lanes going the other, with a central reservation. We negotiated those OK and were now confronted by the same obstacle again. Four lanes one way, four lanes the other.

There were massive overhead traffic-lights controlling the crossroads. When the lights changed, you had to make a dash across to the central reservation, for the final run to the safety of the other side.

The sun was scorching and the entrance to the mall was still a long way across the car park. We made it into the mall and were hit by a blast of freezing air. In an instant, we went from boiling to freezing.

The mall was massive. There were two floors of shops, bars and restaurants. The central focus of the place was a huge ice rink on ground level. You could lean on the upper balcony and watch skaters of all ages.

Near the doors where we had entered was a large steakhouse restaurant. Brewmasters. It boasted that you could drink as much wine as you wanted with your meal. There was a bar inside, so we ordered a couple of beers and confirmed that it was true. As much local wine as you could drink with your meal.

We were definitely eating there tonight. We guessed that we had worn our welcome out with Charlie.

• • • • • • • • • •

It was still hot at eight in the evening. We survived sixteen lanes of traffic and welcomed the air conditioning of the mall. Brewmasters was doing brisk business, but had a table for two and seated us, with glossy menus.

We ordered two T-Bones, medium, and white wine. The wine came in a large, glass jug.

We poured two glasses and tasted it.

"This tastes like water," complained Eddy.

"It is," I said, "At least fifty percent."

"I knew there had to be a catch," Ed grumbled. "I've learned one thing about America.

You can't trust the Americans."

"Excuse me," I said to a passing waiter, handing him the jug, "Do you think we could have some more water with our wine?"

"Certainly, sir," he answered politely, as he removed the jug and also took our glasses. "I'll see to it for you."

"Thank you very much," I said. "Appreciate that." "You're welcome."

He returned smartly with a fresh jug and glasses. He filled them and stood by the table as we tried it.

"How is that, sir?" "Perfect," I said. "Thank you, sir." "You're welcome."

When the steaks arrived, the jug was empty. Without us even asking, the waiter took it away and brought a fresh one. The steaks were huge and beautifully cooked. Halfway through the meal, Eddy ordered another jug. I was about to tell Ed to slow down a bit, when it dawned on me what a waste of breath that would be.

We were having a great time, laughing about the non-existent Sirocco. In a small way, we were disappointed. It would have been fun to own a house in Florida. However, on the bright side, we were having a free holiday on Stephen Hardy and Partners. After a full inquest over another jug of wine, we came to the conclusion that we were Bill and Stephen's loss, not Charlie's.

Hence the reason he seemed so amused when he said, "Over to you, Bill. These guys want to see YOUR Sirocco."

It was getting late. Brewmasters was emptying. We paid the tab and left a decent tip. It was as we stood up that it began to hit us.

We were lagging!

Doing our best to leave with some dignity, we managed to remain upright until we were outside the mall in the air. We managed our way across the car park and stood, staring at the bright lights from sixteen lanes of cars.

"I don't fancy our chances, Ed," I said. "I think we should go back to the mall and call a cab."

We were still weighing up the odds when a voice behind us said, "You ain't gonna make it."

We turned around and saw two American girls laughing at us.

"We've been watching you guys and you ain't gonna make it. Where you headin?" "The Ramada Inn," said Ed, pointing to it. "Just over there."

Laughing also, the second girl said, "I think you guys could use a lift. Stay here. We'll fetch our cars."

They turned to the car park. "Don't even think about it. You wouldn't make it past the first lane."

They weren't long and came back with a couple of medium-size Japanese cars. They both got out and each took one of us by the arm to her car. We both got in the front and soon we were safely outside the hotel. They both helped us out.

"Please will you come in for a drink?" I asked. "You saved our lives. That's the least we can do."

Without a protest, they both agreed. They parked the cars and we led them to a table in *Jack's Place*.

We ordered drinks and sat talking. They were two beautiful, friendly, smart, intelligent girls. They both worked in the mall. The one in a jewellery store that her father owned, the other in a department store. They had just finished work.

They loved our English accents and wanted to know all about us. We gave them a brief summary and told them that tomorrow was our last day. What a shame it was that we hadn't met them earlier. Eddy asked if it would be possible to see them tomorrow night. Had they any plans?

Yes, they had. It was Friday and they always went to a bar called Benigans. Would we like to meet them there?

"Not 'arf," said Ed.

"I take it that means yes," one of them laughed, reaching into her handbag. She pulled out a green book of matches. "Take these. This is the place. Shall we say nine?"

• • • • • • • • •

The cab dropped us outside the bar.

Two doormen scrutinised us. I always favour speaking first. "Good evening."

"Good evening, sir."

"We're meeting two friends. Do we have to pay to get in?" I asked.

"No sir. You're good. We're very careful who we let in, sir. Have a nice evening," he said, standing aside.

"Thank you, too. I'm sure we will."

We found the girls almost immediately. They were drinking what appeared to be orange juice. They seemed delighted to see us and, realising that we hadn't introduced ourselves properly, told us they were Carrie and June.

Carrie was the younger of the two by a few years. She had the clothes, hair and makeup of a supermodel. She seemed besotted with our Ed and was hanging on to every word of his stories. I'd heard them all before.

I was having a great time with June. Although she wasn't in the same grade as Carrie, she was still a pretty stunning girl, with a lovely face and figure. How I wished it wasn't our last night.

I stuck with pints of beer, but Eddy was knocking back different cocktails. He was getting a bit drunk. June told me that Carrie had said in the powder room how much she liked him and wished he wasn't going home tomorrow. She also expressed the same feelings to me. She also wished I wasn't going home.

I made that unanimous.

I took Eddy to one side and said,

"Don't be a prat, Ed. This girl's got the hots for you. Look at her. If you live to be a hundred you're never going to do better than her. Take it easy on the booze. You can get pissed every day of your life, but you'll never get another chance to shag a bird like her. Slow down!"

Well, it's true. You can take a horse to water, but you can't make it drink. I shook my head in despair and, if I'm truthful, a little envy. What a waste!

Your funeral, I thought. Well, I'd stayed sober all night. I hoped it wasn't mine, too. Time to find out.

"I don't suppose you'd consider spending the night with me, would you, June?" I asked hopefully.

"I'm sorry, Des," she said, a little sadly. "I've never done that in my life on a first date."

"I understand fully," I sighed. "I haven't either. It's just that it's my last night and I'll probably never see you again."

"I've thought about that too, Des," said June. "That's exactly the point. I'll never see you again, so it all seems so worthless."

"Worthless to you, maybe, June," I was feeling hurt, "But precious to me. I only know one thing, and that's if I wasn't going back to England tomorrow there's no way that we wouldn't be a serious item. I just know it. All I wanted was an everlasting memory of you to take with me."

"Did you read that horseshit in a book somewhere, Des?" she laughed. "You're in the wrong state. You should be in Hollywood."

"It didn't work there, either," I laughed, "But I meant every word. Honest!"

"I think you could do with giving Eddy a few lessons," she sighed. "Look at him. Totally useless."

"I know," I agreed. "I'm sure he's sorry."

"He will be in the morning. Come on, then, you've talked me into it. Let's get on with it." She had a few words with Carrie. "Carrie said she'll take care of Eddy. I know a motel."

• • • • • • • • • •

We stopped for breakfast at a roadside diner on the way back to the Ramada in the morning. We both said what a great night we'd had, and genuinely wished that things were different. However, the reality of it was that tomorrow we would be on different sides of the Atlantic.

She dropped me off at the Ramada and we kissed our last goodbye. I opened my room to the most awful smell of vomit. It was vile. I opened the balcony doors to let some air in.

Eddy was lying, dressed, on the bed. Fortunately, he'd contained the vomit to the sink and toilet. Some, of course, was on his shirt.

I couldn't leave it for the maid. It wouldn't be fair, and I would be too embarrassed, anyway. I cleaned it all up, rinsing down everything until it smelled a bit better. I got Ed's shirt from him and, despite his protests, binned it.

I told him to get in the shower and pack. We were going home, in case he'd forgotten. We were leaving the hotel about two in the afternoon. I threw all my stuff in my case and checked that I had left nothing. I told him I'd see him in *Jack's Place*.

He joined me about an hour later and told me that Carrie had brought him home. I called him a prat and left it at that. About one thirty, I was pleasantly surprised to see a beaming Charlie walking purposefully towards us. Ed Purser was close behind. Holding out his hand, he said,

"It's been a real pleasure meeting you guys. Ed has the car outside to take you to the airport whenever you're ready. Have a safe journey."

With that he was gone. Finishing our beers, we picked up our bags and followed Ed to the limo. We talked all the way and we all agreed it had been great fun. I for one was really sad to be leaving Florida. I knew I'd be back. It really is the most wonderful place I've ever been.

Checking-in our luggage, we found the bar and waited for our flight to be called. Pan Am were great people to fly with, and it's hard to believe that America let them go bust.

The plane was full. We were in the middle section surrounded by at least a dozen lads, all in the same black T-shirts emblazoned with a white snake. Ed had the window, I had the middle and a long-haired lad had the aisle. He was one of the black-shirt brigade.

They were a rowdy lot, determined to drink the plane dry. As you do, I got talking to my neighbour. They were all roadies for a band, Whitesnake. I'd never heard of them at the time. They were on tour, with concerts in England. They were great guys to be with. The flight turned out to be one of the best we'd ever been on. While half the plane was moaning about them, we just joined in and had a great laugh. They told us all their stories and we swapped a few of our own. David Coverdale was their singer, and they were going to be world-famous.

The flight went far too fast for us and, by the end of it, we had made lifelong friends whom we would never see again. That's how it goes. However, with a personal interest in the band, I kept watch for them and bought one of their albums. It was my kind of music and I went on to become a big fan. I have all their albums, and they always remind me of a great party on a plane with Eddy and Whitesnake.

In London, we all queued up at Passport Control and, after a huge farewell and shaking of hands, we bade each other the best of luck and a good life.

Heathrow brought us back to reality. It was a cold, miserable, wet, dark day. As we boarded our coach to Digbeth, we were pissed off.

Forty

In life you don't always get dealt the best hand. You have to play the cards you're given. Around that time, there was a great song I remember,

"I get knocked down, but I get up again."

It was time to make major adjustments in my life. I explained all to Trevor, who was managing Somerville Road for me, and he fully understood and said he was expecting it. I made him redundant and we parted friends. He found a shop of his own and I did whatever I could to help him.

Leslie Dryhurst, the work experience lad I had taken on and trained, was a very loyal, honest and hard worker. I kept him on full-time and turned him into a very capable butcher. He had a good sense of humour and we worked well together.

I was putting my heart and soul into the shop and it was doing really well. In a small way, I went into wholesale. With all my contacts in the trade, I was getting some amazing deals. I had the shop completely refurbished and I was doing some great deep-freeze packages.

I had a sandwich board made that I put out on the pavement. It was big enough to take four dayglo cards on each side. It was a great form of advertising. I was selling whole lambs, cut and prepared for the freezer, for fifteen pounds each. I could do two pounds of bacon and sausage for a pound.

I couldn't keep up with it. I needed more help. Rita's sister, Sue, knew an ex-butcher who worked in a factory full-time. She said he was looking for a Saturday job. She sent him to see me. We agreed to a trial period. He was a good butcher. He started, and then stayed with me as a friend and employee until the day I retired.

I had landed on my feet.

Ray Jarvis was his name. He was popular with the customers and never let me down. But still I couldn't cope. At that time, there was an insurance salesman for Britannic Assurance who worked all over our area. His name was Colin Kirby. He also was an ex-butcher. He did flexible hours and, after I approached him, he agreed to help me with the preparation of the meat for a few hours each day. He didn't like serving but would come to the rescue if we couldn't cope.

It was a perfect arrangement. Between the four of us, we had it cracked. The great thing was that we all got on and were great friends. It wasn't long before I was having Mondays off, which was great.

The biggest problem of all was having no driving licence. However, there was a good bus service. I got myself a bus pass. From Walmley, the 114 took me to Birmingham Market. If I wasn't going to the market, I got off at the *Bagot Arms* and caught the 28, which dropped me off outside the shop. Tuesday and Thursday, Rita took me to the market and back to the shop.

Another great friend, who delivered all my pork, was Gary Fitsgerald. He used to pick me up Saturday morning at six o'clock and drop me to the shop. He told someone years later that he couldn't understand why he did that.

"There's not another person in the world I would have done that for." Thanks, mate, I'll never forget it.

Around that time there was a huge lamb scare. Chernobyl. A reactor in Russia had exploded and allegedly could contaminate our lamb. A load of bollocks. The price of lamb plummeted. Great. If you could sell it.

Not one to miss an opportunity, I had a poster made for my board.

RADIOACTIVE LAMB

£15.00 EACH

The number of cars that stopped to use the shop was unbelievable.

Eddy and I were playing squash about three times a week, but courts were hard to get at the time we needed them.

I suggested we join the *Albany Club* in Birmingham. It was expensive, but we could always get courts. There were great facilities, too. A swimming pool, sauna and steam room. The works. And a great bar that was always decorated with a variety of nice girls. The hotel was one of the best in town and there was a steady supply of guests using the facilities.

We had to take a medical before they would let us use the squash courts. No problem, we thought. I was declared fit. However, Eddy was immediately banned from the gym and courts. His blood pressure was through the roof. He had had no idea. He had to go to his doctor and the alarm bells were ringing. It was three months before they sorted him out. He would have to take medication for life and cut down on his drinking.

He cut down for a bit and everything seemed OK and we soon forgot all about it.

With everything now in place, I got into the routine I wanted. Three holidays a year.

Bruce the Bacon was another good friend of mine. We got into a routine of going away every Easter. Go Saturday and come back Wednesday.

We tried Cyprus, Florida and Majorca.

In June we had our family holiday to Spain, Crete, or Malta and, on August Bank Holiday, Rita and I would go somewhere in the sun.

It was on one of our family holidays that Ang met a girlfriend in a bar with us. She was American. She was meeting her father for a few days. His name was Ken. He was Commander of the Fleet for the American Navy. The fleet was in Malaga on manoeuvres. We met up in a bar in Torremolinos and had a great few nights.

It was another case of ships that pass in the night. I never expected to see them again.

The following Easter I was in Santa Ponsa with Bruce when we were, surprisingly, in a bar having a great time. There were a lot of American sailors around. Across the room I caught the eye of one of them, staring at me. I stared back in total disbelief. It was Ken, Commander of the Fleet. He was on the town with a group of officers.

Needless to say, it was party time.

It ended with an invitation to join them for dinner in the Officers' Mess the next day. We were to see them on the USS *Puget Sound* in Palma.

The next day Bruce and I got a taxi to Palma. It was a huge port and the grey ships of the American Navy were hard to miss. There were sailors all around and we asked where we might find the Puget Sound. They pointed us to a huge warship at the end of the dock.

There were steps straight up the side of the ship to the deck. A banner on the side announced USS *Puget Sound* Commander of the Sixth Fleet. Two armed sailors stood to attention guarding the stairs.

They asked us to state our business. We did. They radioed and an officer walked down to us. We introduced each other and he pinned badges on to our coats. We followed him up to the deck.

The ship was massive and all around the harbour were warships of all sizes. Ken met us and gave us a tour of the ship, explaining each gun's function and showing us the everyday areas of life on board. He explained that, of course, there were areas of non-admittance.

Checking his watch, he told us it was time to eat. I had a preconceived idea about the Officers' Mess. Tablecloths, napkins and waiters. It was nothing like that. It was a large, steel room with bolted steel tables. There was a long, stainless-steel serving area where you queued for your food and were served as you walked along by cooks in white T-shirts.

I don't remember what I had. We sat at one of many long tables with Ken and other officers. Ken told them about the coincidence of meeting twice and everyone swapped stories of one sort or the other.

After dinner, we took another tour and photos for a souvenir. Just before we left the ship, Ken gave us both two sailors' baseball caps with woven into them.

USS PUGET SOUND AD-38

I keep mine hanging on the wall of my study at home.

• • • • • • • • • •

I was in sight of getting my licence back and started to look for a car. I loved the yellow MG I'd given to Rita but couldn't very well take it off her after three years. Somebody I knew in the car trade was getting divorced. He was ducking and diving here and there, trying to sell stuff for cash that he could hide from the vultures.

He had a mint-condition Mineral-blue MGB GT, with a private plate and a full-length, sliding Webasto sunroof. He had to have pound notes.

I explained the position I was in. After losing my licence for three years, I'd hardly been able to earn a living. The cost of the insurance was probably going to be more than the car. How much did he want? With a sharp intake of breath, I gasped that there was no way I could afford that. My absolute limit was two grand. He moaned that the plate was nearly a grand.

859 BAD. I told him it meant nothing to me, I didn't want it.

I asked him to wait a minute. I disappeared out of the room and returned with two thousand pounds in twenty-pound notes.

The car was mine.

It would be a couple of months before I could drive it, but I think I sat in it every day.

It was the mid-eighties and I was in my late thirties. I had a plan. I didn't want to fall back into the old routine. The pub. Eddy was going out with a girl called Tracy and I was seeing less of him. We let go of our membership at the *Albany* and I was looking for somewhere else.

We were good friends with Henri Taroni, a local scrap merchant, and his family. Rita was always around their farm with his wife, Trixi. Ang was best mates with his daughter, Joanne. Henri loved parties at the farm and one time, when we were boozing in the kitchen, he told me I should join the *Belfry*.

I checked it out. It was everything a spa resort should be. The leisure club had its own bar restaurant by the pool. It had tennis and squash courts, a large snooker room and a great, relaxing, pool area that opened out on to gardens for the summer.

I joined without hesitation. One of the first people I met there was an old friend from Walmley, Jackie.

Jackie was eighteen when we first met. She was a schoolteacher and working part-time at the *Fox*. She was a fit girl and liked to play squash. We played quite regularly but, as happens, lost touch when she stopped working at the *Fox*.

Jackie used the club a lot and we started meeting every Sunday for a swim and a sauna. We used to have tea and toast and lounge around the pool, chatting. She knew everyone, and it wasn't long before she introduced me to Jayne, another schoolteacher.

Soon I was a regular all day on Monday. Then Wednesday afternoon and soon, I was leaving the shop early Tuesday and Thursday.

There were different people there on the afternoons and soon I was friends with a nice woman about my age, Pamela. She apparently was rich and owned nursing homes. Soon, I was meeting her friends. There didn't seem to be many men around. Just all these rich, spoiled women.

Then came along my favourite, Liz. Liz was an outrageous, flamboyant, glamorous and funny show-off. She was married to a radio DJ, Nicky

Steel. It was not unusual to be surrounded by as many as ten glamorous women, drinking tea by the jacuzzi.

They were known as Desie's Angels.

I had to endure endless jokes, as people would say things like, "Are you comfortable there, Des?"

Oh well! I could take it.

With such a lot of idle time on my hands, I had an idea to write a book. It would be just a slim booklet telling housewives all about meat. All the different cuts and how to cook them. I realised, from talking to the Angels, how little the modern housewife knew about meat.

I was also going to explore the area of deep-freeze deals and expose many of the fiddles there were to cheat the customer. I wanted to do it humorously, with cartoon illustrations.

After the *Belfry*, my habit was to go home with the car and walk to the *Fox*. I used to mix with a large crowd of young lads and some of their dads. There were all sorts - students, bricklayers, tradesmen. One lad was an art student, Dominic.

I told him about my book and he was really interested in getting involved. He asked me what illustrations I had in mind. He met me one day in the pub with a folder containing a couple he had done. They were fantastic. Top magazine quality. An art firm would have charged hundreds.

I had another friend in the *Fox*, Rob. His girlfriend was gorgeous. Long, slim legs in skin-tight pants, tiny waist and revealing tops. She was a great, fun-loving girl. She used to flirt outrageously with me. We were talking about my book. She knew Dominic and I told her how good he was.

"I think what you need, Des, is me on the job," she laughed, holding her gin and tonic to large, voluptuous lips that perched below her delicate nose and huge, dark-brown, mixed-race eyes.

"How come?" I asked.

"That's my job," she told me. "I'm a proofreader and typesetter for a publishing company. I could lay your book out professionally. Get me a copy of the manuscript."

I did, and Joan set out some pages as examples. She showed me in the pub. It looked great. "Why don't we meet in the week with Dominic?"

she suggested. "And we can put our heads together."

We swapped phone numbers and arranged to meet at *The Three Tuns* in Sutton mid-week. I got the bus and met Dom and Joan in the bar. It was nice and quiet. We each spread our work across the table and discussed the order of things.

Dominic told us about his friend, Robin Britton, from college, who was in media and all things connected. He wanted to get involved. Dominic said he would bring him along next week.

Joan finished her drink and said she was off. I said, "Me too, I don't want to miss my bus."

"Did you not come in the car, Des?" said Joan. "No, I've only just got my licence back."

"I'll give you a lift. I'm going that way," she laughed. "You can come with me. My pleasure."

• • • • • • • • • •

I was looking forward to our next meeting, to see how the book was going. Joan had laid out some more but needed more artwork. Dominic had another couple of cartoons. Robin was a really nice lad. He did a few jobs on the side in advertising and signs. It turned out that he did a lot of work for the *Belfry*. If they needed a sign for something quickly, they phoned him.

He loved the fun and knew someone who could print it all up in booklet form when it was done. We started to meet twice a week. We needed to, as we seemed to be drinking more and working less.

Soon came the time when it was ready for the printers. Robin took over. The results were fantastic. We did a first run of a thousand copies. They worked out at twenty pence a book.

Amazing.

We met at the *Tuns* to celebrate.

We contacted the *Birmingham Evening Mail* and they did an article with a photo of me.

BUTCHER REVEALS THE TRICKS OF THE TRADE.

It was the book that every housewife should have in her handbag when going to the butchers. It told of the butchers' back slang, a way of talking to each other without the customer understanding. It explained that many unscrupulous butchers were selling thawed-out turkeys as fresh.

Trading Standards said they would investigate.

There was a half-page in the local paper, *The Sutton News*, with details of how to order a copy.

The Mail on Sunday had a half-page in their *You* magazine. Everyone at the *Belfry* wanted a copy. It was all great fun.

• • • • • • • • • •

Eddie's girlfriend, Tracy, was giving him a hard time. She knew about his blood pressure and was always nagging him about his drinking. She was always on his case about his medication, making sure he was taking it. He didn't like being told what to do. One day, he confided in me that he was seeing another girl who was a customer in the shop. Her name was also Tracy.

"Handy if you talk in your sleep," I laughed. "I don't know what to do," he said.

"Well, I'm not really the best person to be asking," I said honestly. "Why not?" he sniggered. "You've had enough practice."

As we chatted about it, I stopped him and said, "Ed, I keep getting confused with which Tracy you're on about. Can we call them Tracy one and Tracy two?"

I told him that the best thing he could do was to keep them both on the go for as long as he could, to give himself time to make up his mind. That was the best thing I could think of and he agreed.

Of course, it was destined to end in tears.

And it did.

• • • • • • • • •

My popularity at the *Belfry* was at an all-time high. The Angels were ever-present and attentive, but it was my best mate Jackie who was always at my side.

I'd stopped taking the MG and used to get a taxi to the club. It was nice to sit by the pool with a pint of beer, even though I wasn't too keen on the plastic glasses. Jackie rarely drank alcohol and had no problem giving me a lift home. Or, should I say, mostly to the *Fox* in Walmley. Sundays we used to meet about nine and, after a swim and a session in the steam room or sauna, we had tea or coffee, with toast. About 11.30 we would get changed and meet in reception or, sometimes, the bar.

Neither of us gave it a second thought. We were just good mates who enjoyed each other's company. So, it came as a great surprise to both of us when, quite casually and innocently, somebody passed some comment, believing that we were a couple. Apparently, everyone had assumed that we were sleeping together.

We both just laughed it off, finding it quite amusing. We'd been friends for a long time and would remain friends even after I met my present wife, Barbara.

As far as we were concerned, we would be mates for ever. It was Jackie who was my rock when Ed fell suddenly ill and eventually died. Thanks, Jackie, if you ever read this.

There was a new kid on the block at the *Belfry*. Brian Simpkins. Brian had just sold his skip hire business for a few million quid and joined the *Belfry*. He drove a new Bentley Turbo R with a private plate, BS 429. He was to become a great friend also. Somehow, we got talking one day and soon were playing squash and tennis on a regular basis.

Jackie was a good tennis player and used to partner me against Brian and his wife, Eileen.

Brian was gaining quite a reputation as a ladies' man. He was always on the hunt.

I wasn't into that sort of thing at the *Belfry* and was quite happy basking in the reputation that I was shagging every girl in the club.

As Brian once said to me, "I suppose that's another one you're not shagging."

Ed wanted to see me one day and I signed him in as a guest at the club. Things had blown up with the Tracy one and two business, and he had decided to stick with Tracy two. I'd liked Tracy one the best, but never said. The worst thing you can ever do is get involved in someone else's love life. You always end up the villain.

It was his business.

• • • • • • • • • •

A lot of people were telling me how enjoyable and amusing my book was. I'd sent a copy to the Book Guild Publishers in Brighton. I had a nice letter back from the boss, Carol Biss, who said what a great idea it was and amusingly written. However, she told me that, as it stood, it was far too insubstantial.

She said that if I could pack it out to forty thousand words then, with the illustrations, she would be prepared to publish it in hardback.

I thought about it long and hard but, as much as I would have liked to, there was no way I could write forty thousand words about meat and keep it amusing.

There's only so much you can say about the subject. However, Carol had planted a seed.

I'd thought about it before, but never really seriously. A situation comedy.

BUTCH

I'd never written a television script and knew that there were many things I would need to learn. It was totally different from writing a book. I did some research and found a correspondence course I could do at home. It would take some time, but I enrolled and got started.

• • • • • • • • • •

Ed decided to get married and set the date for 6th July 1985. He had the ceremony at the Friary Church, Olton and the reception at the *Excelsior Hotel*, by the old airport. It was a grand affair and went off well. Ed had a house in Lode Lane, Solihull. He was working long hours in the shop and going to the pub after work. Then back home.

I wasn't seeing him that much, the main reason being that neither of us would drink and drive and we lived a long way apart.

I was in my routine at the *Belfry*. One afternoon I was sitting with Brian Simpkins, and he mentioned Henri Taroni. I told him that Henri was a family friend. He grinned. I asked him what was so funny? He told me that he and Henri were bitter enemies. Something to do with business. I told him it was nothing to do with me and didn't want to talk about it.

One day Henri invited myself, Rita and Ang to go out for the night with himself, Trixi and the family. There was a new hotel in Walmley, *New Hall*, where many of the stars stayed when they played at the NEC.

It was extremely posh and horrendously expensive, but it was Henri's treat. We were having a port and brandy with a cigar when he mentioned Brian. I told him the same as I had told Brian. Henri was OK with that, but added that as he occasionally used *The Belfry Club*, it would be awkward if they were both there at the same time. I more or less said that we would cross that bridge if we came to it. But I was glad to know where we stood. There were a few close shaves, but we never clashed at the Club.

However, a couple of years later, Henri died suddenly and Trixi was round at my house when Brian called for me to go to the 1990 Motor Show at the NEC. I was terribly embarrassed. However, Brian, acting like a perfect gentleman, strode up to Trixi and held his hand out firmly.

"Mrs Taroni," said Brian, "I'm deeply sorry to hear about your husband. We had our differences, but they were only business."

Trixi shook hands graciously. "Thank you, Mr Simpkins." All was well.

• • • • • • • • •

The London School of Script Writing said that I had a "natural ear for dialogue", which was good. There were ten lessons in the course and, with everything else going on in my life, they were taking time. No matter. My tutor told me there was no time limit. My progress was good. I'd been using my lessons to run with some of the ideas I had for *Butch*. Everything in life was going great.

Then, as always seemed to happen, came the kick in the teeth.

I got to the shop one Friday morning and, before I could even open up, got a frantic phone call. Ed was in hospital. He had collapsed in *The Red House* pub in Solihull at closing time the previous night. He was in the QE in Birmingham. There was no really reliable information, so I went straight up there.

He'd had a brain haemorrhage. He was in and out of consciousness. They were trying to locate the source of the bleed, but there was so much blood on the brain they couldn't find it.

If they could, they would be able to operate.

I was leaning over a lifeless-looking Ed talking to him, hoping he could hear me.

He opened his eyes and said, "Yes, Des?" That was what he always said when he answered the phone. "I've got a splitting headache."

Those were the last words he ever said. He slipped back into unconsciousness again and never came out of it.

It was the beginning of a six-week nightmare.

He was moved into a private ward. All we could do was take it in turns to be with him.

Mom was beside herself and wanted to be there all day and night. Tracy obviously wanted to be there, too. Me and the Old Man had the shops to keep going and could only be there on the nights for a couple of hours. Rita and Ang wanted to visit, as did Tracy's mom and dad.

I found a little sanctuary at the *Belfry*. Jackie was a rock, but realised that you couldn't dwell on it day and night. It was good for me to chill a little with a pot of tea and a natter.

One week dragged into two weeks. It was coming up to the sixth week and it was Ed's birthday on 16th June. Mom never stopped hoping for a recovery, even though the doctors had said it was unlikely.

They couldn't find the first bleed and feared a second one.

Mom insisted on giving him a party, in the ward. I was dead against it. He was just propped up in bed, full of tubes. I couldn't dissuade her and she invited all his friends. I had to go along with it. There was a cake, champagne, snacks and balloons.

Everyone was crammed into the small ward. Ed just sat there, lifeless. I knew that most people were uncomfortable. I thought the whole thing was sick. It felt like a circus.

Four days later, on the 20th, he died.

The ward sister was a friend and customer of mine, Mrs Enright. She had taken me to one side and prepared me.

"Tonight's the night, Desi," she told me. "I'll call you all near the time. Prepare yourself for the rattle. It'll come loud. We call it the death rattle."

I don't know how they knew these things, but she was right, and he breathed his last breath in the early hours of the morning.

There are no words to describe my mother's grief. She had lost her two youngest sons.

She was never the same again.

My own grief could not even be measured against hers. I was just glad that I had all those marvellous memories with him. The one that still makes me smile the most was the Las Vegas hooker.

"You can have one on the house if you like, hon."

• • • • • • • • • •

The dust always settles eventually. After the funeral, everybody had to get on with their lives and grieve in their own way. Mom found hers in a bottle. She was a nightmare to look after. We did what we could and hoped she would come out of it eventually.

I had finished my script-writing course and was working on a sixty-minute pilot episode for Butch. It was called *The Great Turkey Scandal*. I was going to follow that with six thirty-minute episodes. I wanted to test the water, so I sent the pilot to LWT in London.

A couple of weeks later it was returned to me with a covering letter from their Head of Comedy, Robin Carr. He was very encouraging.

"Although I feel this is not for LWT, it is a cut above the dross that usually lands on my desk. My advice is to hang in there and keep writing."

Full of enthusiasm after his positive input, I worked hard for over six months to write the six half-hour episodes.

This time I sent the whole package to Central TV in Birmingham. After a few weeks, I phoned them up. I was put through to Cassandra Smart. She told me it had been passed on to Christopher Walker, the producer. I asked how long it might be before I heard something.

"I don't know," she told me, "But it must have been very good or you would have had it straight back."

After a few more weeks and a few more phone calls, I got it back. They had preferred another comedy that they thought had more longevity. They didn't think *Butch* would be strong enough to run for at least thirteen episodes, which they would need to make it sustainable.

It was also a bit too "close to the knuckle." Also, "Although episode two is very funny indeed, you must curb your urge to shock for the sheer sake of shocking."

As always, I was too far ahead of my time. By today's standards it would seem quite tame.

• • • • • • • • • •

Angela was due to get married four months after Ed had died. 3rd September 1988.

She wanted to call it off out of respect for Eddy. The plans were well advanced.

Everything was booked. There was a new hotel opening in the City Centre, *The Copthorn*.

They were doing us an unbelievable deal, as it was their first wedding. It was really a try-out for them to see how they coped. It could go two ways. It could be a disaster or a huge success. Having spoken to the manageress, I was convinced that they were going to pull all the stops out and make it a success.

I told Angela truthfully that Ed would never have wanted her to cancel her big day. He knew as well as anybody that you have to live life as it comes and live for today. After all,

Angela was marrying the love of her life and, as I said to her, who knows better than us that it can be ended in the blink of an eye?

Thank God it was all a great success. Paul, her husband, was a really great bloke and I and Rita got on well with his family. He had a good job in management at Land Rover and, just coming up to a year after their marriage, on 20th August 1989, they had my first granddaughter, Kimberley.

Of course, she had a huge impact on my life. She was barely out of nappies when I was taking her at least once a week to the *Belfry*. She loved the water and we were always in the pool. She was a huge hit with the

Angels. There was no shortage of girls to look after her when I wanted a swim in the pool, or an hour on the sunbed, or the sauna and steam rooms.

In the summer, we would sit in the gardens or go for a walk in the grounds to watch the rabbits hopping about in the bushes. If I had a game of tennis, she would sit outside the courts in her pushchair and watch. We had great fun.

During all this time I'd been flirting again with my diving. Steve Cantrel, a good friend from the *Bagot*, had learned to dive with Gary Fitsgerald. Gary didn't really take to it much, so I started going with Steve to Stoney Cove on Sundays. We always went early and were back home and up to the *Fox* for opening time.

There was a great gang of lads in the *Fox* and, one day, I came up with the idea of doing an overnight camping trip to Stoney. We organised a few vans and a barbecue and, armed with a load of steaks, chops and breakfast food, we set off on Saturday evening.

I had cleared it with the management at Stoney and they told us we could use the top car park that never got filled until midday. I also informed the Hinckley police, emphasising that there was going to be no drinking and driving involved. Just some grown-up little boys playing.

There were many pubs in Stoney Stanton village. We did a tour and ended up at *The Bull*, the nearest one to our campsite. Coming up to closing, we rushed to the bar to double up our beer.

The landlord looked at us bemused and said, "You're not in Brum now, lads, we won't be closing. There's no rush."

We left at midnight and walked down the hill to the cove. We had formed a circle with the vans, with the barbecue in the middle. Terry, the "Gentle Giant", as we called him, was cook.

He loved doing it and loved even more moaning about it.

Me being a butcher, there were loads of food. Someone else had brought salad and cobs. The dark night was alight from the flames of fat dripping on the coals.

The beer and wine flowed and, when a Panda car with two young policemen arrived, it was hats off and join in the food. They refused a beer but left with bellies full of steak.

Steve and I had sleeping bags in the back of my van. When we woke in the morning it was daylight. We felt a little rough. From somewhere, Steve produced an odd-looking bottle of wine wrapped in straw. It was three quarters empty. Some red plonk he'd bought in Portugal. He shook it and had a swig, then passed it to me. I finished it with a grimace.

I debated with him the wisdom of diving. However, it was all so new to him, he declared, "We haven't come all this way and gone to all this trouble not to have a dive."

It was the start of the end of my diving career. Against my better judgement, we kitted-up and walked down to the water.

It was freezing, and I knew instantly that it had been a huge mistake. We were lucky there was good vis. In deep water, I found myself drifting slowly and happily down. As I looked above me, I could see divers finning near the surface. Everything felt surreal. I was almost on the bottom. I could see no sign of Steve - or was that him floating above me in the green, sun-lanced water? I didn't know.

All I know is that I snapped out of it and went into autopilot. I gently inflated my life-jacket and began to rise slowly to the surface.

I fully inflated my life-jacket and finned on my back to the side. I had just experienced my first and last brush with Narcosis of the Deep. The feeling of carefree euphoria as you drift to your lonely death.

I only had one more dive.

It was a Wednesday afternoon at Stoney. A nice, quiet day. I was with Steve. We went out and down to the bottom. I was having difficulty with my buoyancy. The small air-bottle on my life-jacket seemed to be sticking.

Suddenly, with a loud crack, it jammed open and fully inflated my jacket. I took off like a *Polaris* missile for the surface. Breathing out all the way up, I managed not to burst my lungs. I burst from the surface into the air and landed back on the water with a crash.

Breathing heavily, I took stock of my situation. I was alive. Good. I hadn't felt my lungs burst. Good. I had a tank full of air and, after examining my equipment, felt safe to go back down.

It was important to get down fast, to thirty feet. I was in danger of decompression sickness and needed to decompress. I couldn't see Steve anywhere, so I went down alone. In the shallows on the bottom, I began swimming around leisurely until my air was gone.

There were hard-hat divers walking around in full diving suits, their hoses stretching up to the surface. I hadn't seen them there before and swam about, watching them with interest. As my air gauge registered red, I slowly surfaced.

When I got out, Steve was waiting. "What happened?"

I told him.

He told me that I had apparently wrecked some important experiment the hard-hats were doing. The Stoney Cove people were furious.

"Doesn't your friend know that he can't dive alone?" one shouted at Steve.

"I think he's had an incident and is decompressing," said Steve, in my defence.

"Well, when he comes out, tell him to clear off and never come back again." With that he stormed off.

That day was the decider.

I never dived again.

But I'll always miss it, and never miss a diving programme on TV.

• • • • • • • • • •

It was 1990. Rita, Ang and myself were around at Henri's farm. It was not long after he had died. Trixi, his widow, and some of the kids were there. Peter, Paul, Russel and Joanne. We were having drinks and food.

Joanne and Ang had had this idea to open a shapers' shop. They were the latest craze. Sunbeds, toning tables, nails and beauty treatments. You know the sort of thing. Keeps spoiled, rich housewives out of trouble.

It was agreed that Trixi and I put up half the money each. It was going to be an expensive undertaking. However, with all the lads I knew at the *Fox* in Walmley, I was confident I could get the job done for a fraction of what a shop-fitting firm would want.

I was right.

There was a team of shop-fitters who worked for a huge company and travelled all over the country. Working after work and at weekends, they were ripping through the job. Most of the materials they were managing to salvage from left-overs on jobs.

We were soon ready for all the fitness equipment. A date was set for a grand, well-advertised, champagne opening. Doing the opening was a local celebrity, Rusty Lee. I think her fee was the most expensive part of the job! For what she did, I think I could have done a better job myself.

"Desi and his Angels" open new shapers' shop in Sutton.

It's got a great ring to it!

It was a great success and is still there to this very day. However, for various reasons, there was unrest in the ranks. I think the long hours and family commitments proved to be too much.

We all had a meeting at the farm and it was agreed that Ang would sell her half to Joanne and retire to a life of leisure.

Happy days.

Forty-one

It was around this time that I was reading my paper on a lounger by the pool when I spotted a familiar woman with a girlfriend on loungers across the pool.

It was Lyn Sassons, wife of Chris, who had been in prison. We spotted each other and ended up having tea and coffee. Her friend was Barbara Dawkins. They were new members. We talked a little about old times and parted with the usual,

"See you around, then."

From time to time I would see them around the pool. We were just on nodding terms. They used to meet for a game of tennis. One day, Lyn asked me if I played. I said yes, but not often. She asked if I'd like to join them for a knockabout sometime.

Spring, summer and autumn came and went. I used to take Kimberley swimming every Wednesday. The Angels were always there, fussing around and it was all good fun.

It was too cold to play tennis. One day I was having coffee with Lyn and Barbara when I suggested that, as it was no longer tennis time, perhaps we could try some indoor sports.

Why not go for a drink somewhere? Barbara said that she couldn't possibly do that, as she was a married woman with two kids. I thought, "Fair enough," then Lyn piped up with, "Why the hell not? What's the difference between having a game of tennis and having a drink?"

Over the winter, I started meeting Barbara on a Monday morning. We would go for a drive out somewhere and have lunch. Then, as the spring began again, we arranged to play tennis again. However, we both found that we preferred the day out with a few drinks and lunch.

We found ourselves drawn to Derbyshire. We found some nice places in Melbourne and Appleby Magna. Ashby de la Zouch was also a nice place.

Quite by coincidence, I was holidaying in Florida that year with Rita. Barbara and her husband and children were also going to Florida at the same time. Rita and myself were going to Countryside, where I had been with Ed. Barbara was doing Disney World. Not my scene.

It had never been my intention to get too involved with Barbara. I didn't want the complications. However, with these things, at some point it begins to get out of control. I realised while I was in Florida that I was missing her. I kept thinking how odd it was that we were so far from home, yet she was not that far away. I found myself wondering what she was doing.

When we got back home, we resumed our Monday outings and soon found ourselves fitting in the odd night out. However, you still could by no means call it an affair. We weren't committed to each other in any way. We both had our families. As I always used to say, "Just as friends, nothing in it!"

We were still seeing each other at the club and, sometimes, Barb would play with Kimberley, who was about three by now.

I made friends with a new girl at the club. Samantha. She was very young, about twenty, and was a student. She used to sit with myself and Jackie and Brian Simpkins, who found it all very amusing.

One day the Angels were out in force. I was with Kimberley and surrounded by beautiful girls. We were all taking turns to watch Kimberley while we used the sauna, jacuzzi or steam rooms. Jackie and Samantha used to like to do a good few lengths. Mostly breaststroke.

As we were sitting around the pool having fun, I confessed that I could never master the breaststroke. Samantha thought that was very funny.

"I'll teach you, Des," she offered. "It's easy. Nothing to it."

Thinking it might be fun, I agreed. We arranged a day for my lessons and I met her by the pool. We were sitting with Brian, when Sam announced that it was time for my first lesson.

Well, talk about a laugh. It was hilarious. Sam and I were thrashing about in the water, arms everywhere. She was trying to support me while I was trying to swim like a frog. I was sinking. Sam was saving me. We had a very amused audience as all my efforts failed and Sam dragged me to the side, squealing with laughter.

We climbed out, much to everyone's amusement.

"That was fun," gasped Sam. "You nearly had it. We'll get there next week."

I agreed, and my next lesson was booked. Brian was chuckling away behind his newspaper.

"How is it, Des, that a scuba diver with your experience can't swim?" "Diving's completely different from swimming, Brian," I told him. "Yes, Des. I'm sure it is."

Sam's boyfriend lived on a small farm in Walmley. She used to go there after the *Belfry*. I was still taking a taxi to the club in case I wanted to have a drink. If, for some reason, Jackie couldn't take me home, I got a taxi back. It was cheaper than a barrister.

However, as Sam was also going my way she would often take me. On the way home we used to pass a really old country pub, *The Cock*. They did an early bird menu, which was amazingly cheap. I started taking Sam there as a treat for giving me a lift. It wasn't much dearer than the taxi fare. Sam was great company and it was no hardship driving with her.

On one of my early-evening visits, a new waitress had started. It was Bob Collard's daughter, one of my mates from the *Fox*. She was only about eighteen and seemed shocked to see us. I wasn't bothered, Sam was just a friend.

However, she spread the gossip like a wild fire. She told her mom and dad. By telling Bob, you might as well have put an announcement in the *Birmingham Mail*. He was known as the Duke of Walmley. Soon everybody in Sutton knew how his daughter, Donna, was serving Des and his teenage girlfriend every week at *The Cock* at Wishaw.

I found it quite amusing and, of course, denied everything. With a smile. Of course, there was also the inevitable gossip at the *Belfry*.

Brian said from behind his paper,

"There's a rumour going around that you're giving Sam one." "No!" I said, aghast. "Who started that?"

"You, by all accounts," he chuckled.

Brian could hardly talk. He was gaining quite a reputation as a serious player. He was known, unlike me, to have had a series of affairs. In fact, I knew him to have a full-time young mistress, Angela, who had been his secretary. He had set her up in a flat and looked after her. He confessed to me once that he was quite serious about her but had too much to lose to leave home for her.

Even with a wife and two children and a mistress, it still wasn't enough for him. He still liked a bit on the side. One Monday afternoon he was on heat over a very attractive female member who was "up for it."

"Des," he whispered, "Have you got any cash on you?"

I always had a lot of cash with me, the shop takings. "Sure," I said, "How much do you want?"

"A couple of hundred, if you've got it. I'll give it you back on Wednesday," he promised. "I want to take Kath to a hotel, for the evening. I don't want to risk using a credit card."

"No problem," I told him. "I'll slip it over to you."

I kept my wallet in my sports bag by my lounger. I sorted out some cash and passed it to him.

"Thanks, Des, you've saved my life."

"Have fun," I grinned. I found it quite amusing to be lending money to one of the richest blokes in the club.

I got it back, of course, on Wednesday, along with a pint of beer as interest.

The year was slipping by again and I found myself seeing more and more of Barbara. She was made a member of the Angels and happily joined in all the banter with Sam, Jackie and Liz. For some reason, she didn't like Liz. Liz was very outrageous, she loved to shock and be the centre of attention.

It didn't help when one afternoon she cat-walked from the changing rooms in a white, transparent swimsuit and sat down next to me. There was quite a gathering of the Angels that day. Liz was full of herself and proudly thrust out her chest.

"Des," she said, "What do you think of my new tits? I've always wanted bigger ones, so I've been out and bought some."

To everyone's amusement, I chuckled and inspected them closely. To be fair, they looked great and I told her.

Dramatically, as if on cue, she said, "Would you like to feel them, Des?"

Do I like cocktails?

With a great deal of seriousness, I began a detailed and lengthy inspection. I cupped them in my hand to size them and lifted them up to weigh them. Gently wobbling them for natural bounce, I was giving them my full approval.

"Be careful with them, Des," Liz warned. "They're still a bit sore." Trying to get in on the action, Brian asked, "Can I feel them, Liz?"

In true Hollywood fashion she defended them with cupped hands and cried, "No. They're only for Des."

By now everyone was falling about laughing. Pamela, one of the Angels, thrust her backside in my face and cried, "What do you think of my bum, Des?"

Happy to fall in with the fun, I explored it carefully and declared that I thought it was fine.

Just great.

"Anyone else want an opinion?" I asked hopefully.

Barbara and I still laugh about it to this day. Me slightly more than her, I suspect. It's still one of my fondest memories of the *Belfry*.

Inevitably, it was bound to happen. We were falling in love.

We always wanted to see more of each other. I booked a room one Sunday night at the *Belfry*. I don't know what story I told Rita. We were going to have a meal in the French Restaurant. I think Barbara told her husband that she was having a meal with the girls.

I got to the hotel late afternoon and checked in. The receptionist was a friend of mine and told me that the hotel was very quiet that night. She had arranged for me to have the Balasteros Suite, one of the best in the hotel. It overlooked the 18th green and was accessed by a metal, spiral staircase.

Very posh.

I met Barbara on the night in the restaurant. Her husband, Peter, had given her a lift so she could have a drink. Not great, I thought. We had a nice meal and opened a bottle of champagne.

What happened next is still a little vague in my mind. It was pre mobile phones in those days. Barbara had to go and make a phone call. When she came back she was in a panic. She took a sip of champagne and said she had to go. Peter was picking her up.

With a half-garbled explanation she dashed off, leaving me alone in the French Restaurant with a bottle of champagne. I ended up in the Balasteros Suite alone, with a bottle of champagne, watching TV in a four-poster bed.

Later, the room phone rang. It was Barbara. Something had gone wrong, she would explain in the morning. She was going to get out of the house early and come to the suite as soon as she could.

Great!

I made use of room service and had a few complimentary drinks courtesy of my friend on Reception. The suite was luxurious and I slept well until there was a knock on the door about 8 o'clock the next morning.

It was Barbara.

Little did we know that the shit was soon to hit the fan. Very soon.

Forty-two

It was about nine in the morning. I was leaving the main entrance of the hotel with Barbara, when I saw a sight that made my blood run cold.

Rita and Peter were storming towards us across the car park. They had put their heads together and were staking us out in their cars. One was watching the hotel entrance, the other the Leisure Club entrance.

We panicked and made a run for it, to our separate cars. What happened next is vague in my mind. It was very Keystone Cops. Rita was chasing me in her car and Peter was chasing Barbara in her car. Racing through the country lanes, we all ended back at our respective homes.

A few hours and many phone calls later, it was decided to have a showdown at Barbara's house. We met in the front room to try and sort things out.

I realised that discussions weren't going well when, out of the blue, the palm of Peter's hand flashed up and struck the underside of my nose with an upward thrust. Blood spurted like a fountain from it and I crashed backwards into the TV and stand in the front window.

The stand smashed and I lay on the floor, covered in blood and wood from a blow that I later learned was a shot aimed to kill. It was meant to push the nose up into the brain and cause instant death. As an experienced karate instructor he obviously knew this, and could have killed me.

Not exactly Queensberry Rules.

Rita was as shocked as anyone. She never expected it to be like that. She helped me up out of the debris and, with blood soaking from my nose down to my favourite T-shirt, we decided that the summit was over.

The next few months were mad. One minute it was over, then it was back on. We moved out into a flat together, then we moved out again. Those days were so confusing that, subconsciously, I've blocked them from my memory.

Amnesia.

I have no memory of them. And don't want to.

I went on holiday with Rita and Kimberley to try and find some peace and answers.

It always came back to one thing.

I loved Barbara and wanted to be with her. End of.

My good friend Johnny Sabine, from the *Fox*, whom I had helped through his hour of need, came up with an idea I liked.

"Desi," he told me, "You need to get away from both of them and clear your head. I'm looking to buy a place in Lanzarote. Come with me for a week and I'll try and sort you out."

So that's what we did. John booked it all and we had a lovely apartment by the pool in Porta Del Carmen.

It was heaven. Peace and tranquillity.

We were sitting at the bar in the Old Town one night, when two beautiful Scandinavian girls sat by us. The one introduced herself as she sipped a cocktail. In a very pronounced accent she said, "Ve are Norvegan."

"Piss off," said John, rudely.

"Hang on, John," I said, "That was a bit hasty."

"And you can shut it, too," he said angrily. "You're in enough trouble. We've come here to sort you out. Not make it worse."

"But I only thought…" I started.

"Yes, I know what you thought. You can forget it." "But John…"

"Bollocks." That was that.

I don't think he thought much of my idea for a solution. Shame.

• • • • • • • • • •

I don't think the holiday with John achieved much more than a week away from it all, and a bit of much-needed fun.

The situation was still the same when I got back. It looked like the only solution was a divorce.

Divorce, I concluded, was probably worse than death. With death, it's all over and you know nothing about it. Divorce is agonising pain that even morphine cannot end. Not only do you suffer, but your family and friends suffer, too.

People are divided in their opinions and try to avoid you. Children's lives are torn apart through no fault of their own.

The person you loved and lived with for twenty years becomes your sworn enemy through no fault of her own.

It's survival.

Neither of you wants to end up with nothing. Back to where you started. The knives come out and no quarter is given. Overpaid solicitors ring out the battle cry and soon you're in court fighting it out down to the last penny.

With an execution, the waiting is the worst part. The end, when it comes, is swift and permanent.

Not so with divorce.

Divorce is not the end, but the beginning. For the children, life becomes a tug-of-war, pulled from one parent to the other. It's a bungee jump. Down into the depths, then snatched back up again. Time may heal a deep wound, but the scar is always there. Like a tattoo, once you have it, you can never get rid of it.

Then there's the baggage that comes with it. Barbara has two children. They were young at the time - Anthony was about ten, Natalie was about thirteen. They would enter the equation later.

Fortunately, Angela was grown-up and married, with Kimberley, but her life was still badly affected by the divorce.

The whole thing is a bad, messy business, as anyone who's been through it will identify with. The sad thing in my case was that I didn't hate Rita. I'd met Barbara and, through no fault of mine, we'd fallen in love.

As I discovered later, Rita didn't hate me either. She just wanted what she thought she was entitled to. I didn't want to give her any more than I had to. Simple as.

When the dust all settled, Rita had the house lock, stock and barrel, her nice MG, all the furniture and a few quid.

I had a few quid and my business, and nowhere to live. My dad was living alone in his spacious, three-bed semi in Sheldon and offered to take us in. It was a welcome sanctuary. In his mid-seventies, he was still active but not as confident as he used to be. He had a brother and family in Guernsey and wanted to see them one last time. As a way of thanking him, we took him there for a week and stayed at his favourite hotel. We hired a car and took him visiting and anywhere he wanted.

It was what we all needed. We formed a strong bond that was to be unshakeable. Dad loved Barbara and she loved him.

Things were finally starting to sort themselves out.

Now I had to get going again and get back the ground I had lost. I still wanted to retire at fifty.

Forty-three

Retire at fifty. Some hopes. I was forty-four. A big ask.

The shop I was in at Somerville Road had taken a bad hit when Asda had opened a huge store half a mile away. I had my eye on a promising position in a block of shops not too far away, in Hobmore Road. It was a busy area.

I signed a six-year lease that would take me to my fiftieth birthday. I rented Somerville Road to some Indians to do their Halal chicken. My existing customers were close enough to the new shop for them to follow me, so I had a firm customer-base to start with.

I fitted out the shop with most of the equipment and refrigeration from Somerville Road and opened up. It was a great success; with my contacts in wholesale, I was also supplying many butchers I knew with some good deals. I was making money again and stashing it.

We were getting regular again at the *Belfry* and collected Kimberley every Wednesday afternoon and took her swimming. It was great. She and Barbara got on well. We never missed a week.

The Angels were disbanded, but there were still a few survivors. Jackie, of course, was one. We still met Sundays for a swim and tea and toast.

Samantha broke up with her boyfriend in Walmley and disappeared from the scene. We sometimes spoke on the phone. She was happy and had moved on. Liz was in a serious relationship and never spoke if he was around. I think she was always fearful that the exploration of her plastic enhancements might come to light.

Ray, my Saturday butcher, was still with me, so we managed to get our annual holidays. Cyprus and Malta were favourites. Colin the insurance man was still helping a couple or three days a week and everything was going smoothly.

The years were drifting by and soon my lease was nearly up. I had decided I was going. I was 49. It was April. There were still six months left on the lease. My accountant advised me to close before 6th April, the start of a new tax year.

I had fallen out with the landlord some time previously. He was collecting his rent in cash every Tuesday. On Saturday 4th April, I brought down the shutters for the last time. On Sunday morning Curley Bob, my mate, brought his Warwick Pork van to the back of the shop and stripped the place bare.

The only thing left in there was the large cold-room. I had sold that to County Refrigeration and, on Monday morning, they dismantled it and took it away.

I never set foot in the place again.

· · · · · · · · · ·

The first thing I did on Tuesday morning was go down the market. I had, of course, told all my suppliers that I was retiring. To many, a retirement usually meant bankruptcy. It was popular among butchers to run up big bills with their wholesalers and then go bust, taking with them a few thousand quid that they owed them.

Not my style.

With a pocket full of cheques and cash, I paid everyone every penny that I owed them.

Respect and my good name were worth more to me than a few grand. To be fair, nobody expected any less from me. I could show my face anywhere down at the market without skulking around avoiding people.

I bumped into a butcher, Frank Kenard, who had a shop in our area. He greeted me like a long-lost friend. "Des! I hear you've retired."

"Yes, Frank and, before you ask, I'm not doing any holiday relief," I told him firmly. "I didn't like running a shop of my own, so I'm certainly not running someone else's."

He looked deflated.

In those days, most butchers found it a nightmare trying to find someone to manage their shop while they went on holiday. Most relief butchers were cowboys, who weren't up to the job and only wanted to line their own pockets. Horror stories were well-documented in the trade. When a good, honest butcher was available, he could name his own price and be constantly in demand.

Not for me. "Sorry, Frank."

I was, however, amazed by the number of my old customers who contacted me to get them meat. Butchers were becoming like snow in July. There was none. My customers had been spoiled. They could not get what they wanted from Asda.

I started supplying them and found that I was really enjoying it. I could still keep in contact with all the lads at the market and, at the same time, make a little beer-money.

Happy days.

I was in need of a decent holiday.

My thoughts drifted back to when I had "lived in Spain" for a month.

I thought how good it would be to "live in America." That was my plan. To "live in America." Not California and not Miami, nor all the tourist areas like Clearwater and Tampa. Certainly not a shit-hole place like Orlando.

I started my search.

I saw an advert in one of the daily papers. Some bloke in England owned a house in Spring Hill, Florida. Spring Hill was a small, one-horse town about a hundred miles north of Clearwater on the Gulf of Mexico.

It was a two-bedroom, two-bath spacious bungalow with a private pool on a large plot in a quiet, residential area. Not a tourist for a hundred miles south and a thousand miles north. There was a shopping precinct with supermarket, bars and all the normal usual shops. Even a couple of thrift shops (charity shops to us).

It sounded great.

The rent worked out at five hundred quid for the four weeks. That included all utilities, a gardener and a pool-cleaner. We booked it for September, Barbara's fortieth birthday. I booked two return flights direct to Tampa from Heathrow. I arranged car hire from this end to save any hassle when we got there. I wanted something a little better than a Ford Fiesta, as I had had on my previous trips. I chose a Chrysler Neon, with full Air. I knew from experience how debilitating the heat could be in Florida.

As a thank-you to Ray, my Saturday helper and friend of many years, I told him about the trip. I said that if he and his wife, Pat, would like a week in Florida for nothing, all they had to do was book their flights and I would pick them up at the airport and they could stay at the house for nothing. They jumped at the chance.

We were good to go.

· · · · · · · · · ·

Tampa International is a vast, bright, spotless airport. Immigration was friendly and helpful. We were soon in our Chrysler, studying maps provided. We found Spring Hill. It looked fairly straightforward, as are most of America's freeways. There was a toll road we could take and were advised to do so. Good move. It was traffic-free and took us most of the way.

Following instructions, we found ourselves at the bottom of the drive of a beautiful American home with immaculate, large, grass frontage. Holding the keys we had been posted, I pressed a fob and the garage door began to slowly rise. It was a large, two-car area with a huge washing-machine and clothes dryer. We drove inside, pressed the fob and the doors closed behind us. A door from the garage linked us to the house.

We went exploring. The house was a dream. Huge lounge, beautifully furnished, with American fridge-freezer with ice-maker. There was the usual starter pack of basic food and, in the cupboards, tins of food left by previous renters or the owners. There were even some beers in the fridge. I took one out immediately and drank it.

The two large bedrooms were on either side of the house, with their own private bathrooms.

A door from the one bedroom opened out onto the swimming pool, which was completely enclosed in a mesh area to keep out bugs. A door from there opened onto a large patio and well-groomed garden. We loved it!

With duty-frees and some beers in the fridge, we decided to stay "home" and have a night in watching TV on the hundreds of channels available.

It was our first day living in America. We went shopping. The local supermarket was like a massive Sainsbury's. The fresh produce was amazing. The displays of fruit and veg were so immaculate you didn't want to touch them. I was drawn to the meat counter.

Its packaging, preparation and variety were light years in advance of ours. I was mesmerised and was buying trays of flavoured pork ribs and chops. The steak on display looked mouth- watering, especially my favourite T-bone.

Next, I moved on to the beer, wine and spirit aisles. Apart from Bud and Micky, I recognised none of the others. There was one that caught my eye, Old Milwaukee Lite. It was on offer.

Twenty cans for $8.99 - just over a fiver! I checked the proof. 5%. Good. I tried a slab. There was a vast selection of wine. Californian was the cheapest. They did gallon jugs of Chablis for under ten dollars. I had two of those. One white. One rosé.

The tobacco counter sold White Owls, a real favourite of mine, virtually unobtainable in the UK. I bought a box of fifty for practically nothing. Fully stocked with all the essentials for survival - beer, wine, cigars and some food - we went back "home" to put it all away.

The other essential for community life is, of course, a friendly local bar with a happy, smiling waitress. I made my mission for the afternoon to find one. Dressed in T's, shorts and trainers, I decided to drive in search.

We took the main road south to Clearwater. There were four busy lanes in both directions. On either side were car dealerships, used-car lots, eating venues and dozens of sleazy-looking places with bright neon signs advertising strip-shows and all-nude girls. The usual garden centres and furniture stores littered the place, but nothing I could see that resembled a nice, waterfront restaurant bar.

I was beginning to wish I had Ed Purser by my side to find me a waterfront bar in a nice marina, while he went home to do his washing. We'd driven far enough, and I decided to turn around and head back. Also, I thought it would be a good idea to teach Barbara how to drive in America. We agreed she should have some practice early next morning around the deserted streets by our house.

We were about halfway home when I saw a little sign by a narrow road on the left. *Sam's Bar.* I pulled a left turn and followed the road through a small, residential estate of bungalows, condominiums and mobile homes. It ended in a huge, open area overlooking the ocean. There was a large car-park and a tropical theme bar with a huge, open-air patio with a stage area and dance floor.

Small boats and jet skis cruised the waters and disappeared in and out of canals that threaded like roads through the community. There was a sparse sprinkling of people, mostly retired I would say, fishing or walking, or just sitting drinking-in the luxurious views of the Gulf of Mexico.

Herons sat on posts in the sea. Gulls and other seabirds hovered and glided in the air, taking advantage of the warm and cool thermoclines to help them travel effortlessly. There were only a few cars in the car park; some were pristine classic, 1950s and '60s convertibles. We had stumbled on a well-hidden local retirement community.

Everyone was friendly and ready to pass the time of day. We walked about for a while and took a table in *Sam's Bar* overlooking the ocean. The waitress was young, slim, friendly and comfortably dressed in bikini top and cut-off shorts.

"Hi. How you guys doin'? What can I get you? Today's special is clam chowder with hot bread rolls," she said, order book in hand and pen at the ready.

"Beer and a large house wine, please, "I answered.

"Bottle or pitcher?" she asked. "Pitcher's best bet if you're planning on stopping a while." "I'd have a pitcher," I told her, "But it gets too warm in the sun."

"We got that one covered, hon," she smiled and bounced off.

She came back shortly with Barbara's wine and a pitcher of beer and a glass that she set down on the table for me. Hanging on the side of the jug was a long tube of ice cubes. She left it and cheerily walked off. I poured a glass and had a drink, then popped an ice cube in the glass.

"Don't think much of this idea," I complained to Barbara, "It just waters down the beer." "I'm sure you'll manage to drink it before it gets too watery," she said, with what I thought was a slight note of sarcasm.

"I'll try."

The waitress reappeared shortly at my side. She leaned in pleasantly close to me and took my tube of ice. She hung it inside the jug.

"I think you'll find it works better inside the beer, hon," she said, and laughed quietly as she took her long, tanned legs back to the bar.

Chuckling quietly to myself, I thought,

"You prat."

• • • • • • • • •

Back home, I lit the gas on the barbecue, which was under the screen by the pool. It was a gloriously hot evening so, before I started cooking, I slipped out of my shorts and went skinny-dipping in the pool. It was 100% private, so there was no point in soaking my shorts. Barbara was on a sun-lounger in the garden reading a book.

Refreshed from my swim, I towelled off and started cooking. T-bone for me, sirloin for Barbara. The sliding patio doors had to be kept closed at all times for the air con. It was wonderful to pass from the heat of the garden to the cool of the house.

The Old Milwaukee was great. At five quid a slab it was the beer of the day. The Californian Chablis was probably one of the tastiest wines I'll ever drink. I am definitely no connoisseur, but I do know if I like it or if I don't. Standing at the barbecue cooking, with a glass in one hand and a fork in the other and a White Owl wedged between my teeth, I decided that I wanted to live in America.

Once again, I was smitten by the place.

We settled in like true natives and soon the first week was gone and I was preparing for the arrival of Ray and Pat. We were planning to collect them from the airport and had shopped well for all the essentials.

By the way, Barbara had mastered her driving after a difficult start. Her first lesson had been a total disaster and I thought I was doomed to be the sole driver.

Trying to cope with the controls being on the left-hand side was bad enough, but the car being automatic was even worse. Driving on the wrong side of the road made it even more scary. But, all of a sudden, it just seemed to all fall into place as she got the hang of it. She was so excited and I was so happy for her, that I told her she could drive all the time if she liked.

"Honest, Des?!" she exclaimed. "Thanks."

"No problem," I said, leaning back with my beer.

It was time for the airport run. It was about an hour and a half drive. We collected our house guests, to the usual cries of delight and excitement. Arriving back home, they loved it. I'd cooked a beautiful piece of beef with mash and fresh veg and all the trimmings. We spent the evening drinking and laughing and planning what to do on the days they would be here.

The main thing they wanted to do was Orlando. Disney World. Shit!

I hate theme parks.

The first day we stayed local. I took them to the shops so that they could buy whatever supplies they wanted. On the way back, we stopped at *Sam's Bar* and had a few drinks between exploring the canals and ocean front. There were picnic areas and toilets and a couple of piers to fish from.

On one of the piers there was a small, excited group of people pointing out to sea and sharing a pair of binoculars.

"Look at those guys feed," exclaimed an old guy in a sailor's hat. "Watch 'em go. Take a look."

He handed me the binoculars and I trained them on an area of thrashing, foaming water.

There was a glimpse of a pointed head and a black triangle. Two sharks were feeding.

"Found a mass of fish, I guess," said the brown, weathered old-timer. "Always fascinates me. Don't matter how many times I see it."

I agreed and passed him back his glasses. I loved this small, secret part of Florida. I learned that the locals didn't mind tourists. They just didn't encourage them.

"A few's OK," I was told. "We like a few. Just not too many."

A week's not long in Florida. We took Ray and Pat to the usual tourist places. Clearwater Beach, Treasure Island and a Greek community that had made a living for years diving for pearls.

We took them to Disney World, which took a full day. We got lost on the way and ended up in downtown Orlando. Not a good place to be lost.

Barbara saw two black youths sitting on a wall and stopped the car. "What are you doing?" I asked, astonished.

"I'm going to ask them for directions," she told me.

"Are you bleedin' mad?" I shouted. "Get the hell out of here. I'd rather drive around all day than get out of the car here."

To be honest, the visitors were starting to get on my nerves. They did nothing in the house and just expected meals to appear on the table. They were expecting to be driven anywhere at any time and, when I stopped to fill up the car, they never offered a cent towards the petrol in the free car they were being chauffeured around in.

Taking the piss, or what?

There had been stories in the news all week about Hurricane George, which was sweeping across the Caribbean. It could possibly be heading our way. We were put on high alert.

Ray and Pat had two days left. Well, one, really. The last day would be spent getting them to the airport for their flight. There were two things left that they wanted to do. One was to go to the mall at Countryside, dropping in at a large outlet park on the way to do some shopping.

The second thing was they wanted to go to a highly-acclaimed restaurant, *The Black Otter*, which was nearly an hour's drive from the house. That would mean Barbara driving all day to Countryside and back and then driving all night to *The Black Otter*.

I gave them an ultimatum. Countryside or *The Black Otter*. Not both.

They decided on the shopping. That suited us. We didn't want to be driving miles at night with Barbara unable to have a drink.

We got to the mall OK. I wasn't interested in shopping. I bought a massive newspaper, that you could have lit twenty fires with, and settled down at the bar in the mall. Barbara, of course, couldn't resist the shops.

I couldn't wait to see what thank-you gifts Ray and Pat would buy us. Perhaps a gallon of petrol or a jar of coffee! I told them I was going nowhere and, when they were done, to meet me back here.

When they did get back, Ray looked either pissed or on drugs. I asked Barbara if they had been for a drink. She said no. However, now that she thought about it, she sometimes suspected that Ray had a bottle in his bag. I knew he was partial to Scotch.

Ray, Pat and Barbara joined me at the bar and I bought some drinks, that is to say, I put them on the tab. Ray ordered more, on the tab. It was time to settle up. I took the bill and asked Ray for his share. He started questioning it and asked to examine the tab. The female bar-keeper had obviously seen it all before. He was embarrassing me, so I told him to give me the bill and I'd settle it. We could sort it out later.

He didn't give me the tab. Instead, he told the bar-keeper that he needed time to check it all out.

"You do whatever you gotta do," she told him. "I'll be right back." Seething, I snatched the check from him and paid with a generous tip.

"He's pissed," I said to Barbara. "I don't know what he's been drinking, but it's not what he had in here."

It rained stair rods on the way back. I was pissed off. The roads were turning to rivers. Barbara had to pull over as the wipers couldn't push the rain away fast enough. Huge lorries washed past us, drowning the car in waves of water.

It was scary.

Ray and Pat were in the back. Ray was passed out. I told Barbara to get going. It was safer on the road than the hard shoulder. We were nearly home. When we got to Spring Hill, it was like it had never rained. The sky was clear and the sun was shining. The roads were dry. It hadn't even rained there. Spooky.

Back at the house, Pat got a glass of red wine without a care in the world, blissfully telling us how much she was looking forward to *The Black Otter* that night.

Apparently, it was going to be their treat. They were paying.

Barbara was torn. She said she felt guilty about it and didn't mind driving. I told her, "Over my dead body". I poured her a large glass of Chablis and told her to drink it.

"I can't," she said. "Not if I'm driving."

"You're driving nowhere," I told her angrily. "Now drink it! She's sitting out there with a book and a bottle of red wine and expects you to go all day and night without a drink so you can take her to *The Black Otter*. Well she can get stuffed."

Ray was getting comfortable with a book and a beer, when I told him,

"I hope you don't think we're still going to *The Black Otter* tonight, mate. I told you this morning that it was one or the other."

"But Pat's really been looking forward to it," he protested.

"Well, Barbara's knackered and having a drink. If you really want to go that much, you can call a taxi."

Well, that was the death of the holiday. I knew it would be. Pat disappeared into her room and wasn't seen for hours. Ray kept hovering around. Pat had always been the boss and, when she didn't get her way, she made his life unbearable.

"Is Pat all right?" Barbara asked. "Only we haven't seen her all night."

"She's not very well," he said, by way of an excuse. "She's got a bad headache." "Well, far be it from me to make it worse," I told him, "But have you seen the news?

Hurricane George is heading our way and they're starting to evacuate the area. They're advising people to board up their houses and evacuate."

"I've seen that, yes. So, what does that mean?" he asked.

I laid it on the line. "It means that, if you want to be sure of getting your flight home, you'll have to leave in the morning."

"But that will be a day early," he protested. "Where will we stay?"

"A hotel," I said simply. "The fact is that if this hurricane comes here then there's no way Barbara's driving you to the airport in the middle of it. You can go tomorrow before it gets here, or you can take your chances. You'll miss your flight if it hits us. It's up to you. I don't mind either way. But, if you want to be sure of catching your flight, you'll have to go in the morning. Don't forget, you've only got to get to the airport. We have to get back!"

I could see he didn't like it, but they'd pissed me off and I was getting rid of them. "Think about it and tell me what you want to do."

He disappeared for an hour into the bedroom with Pat. When they re-emerged they put a brave face on it. They had decided to go in the morning.

The trip to Tampa was a quiet affair. There were hardly any cars going our way. Most of the traffic was streaming the opposite way. North, away from the oncoming hurricane. The skies were dark and angry. Miles ahead the sky lit up from lightning. The rain was heavy but not bad. Barbara drove well. We took the toll road. Ray didn't offer to pay.

At the airport we exchanged kisses and goodbyes, wished each other luck and they promised to phone when they were settled in their hotel. We turned back to Spring Hill.

"They won't phone," I said to Barbara.

"Of course they will," she said, brushing aside my cynicism. "They won't phone."

"They will." "They won't." They didn't.

They didn't for the next twenty years.

I never heard from them again.

The moral of the story?

If you want to fall out with your best friend, treat them to a free holiday in Florida, buy them food and drink for a week, drive them all around the State and risk your life and limb to get them to the airport before the hurricane, and you can't go wrong.

Simple.

On the way back, we stopped at the supermarket for supplies. People were panic-buying bottled water. Pallets of it were being dropped in the aisles and snapped up. We stocked up with Old Milwaukee Lite and Californian Chablis. When we got to the checkout the friendly girl said, "I can see what you guys got planned for the hurricane."

She passed us a leaflet on how to survive a hurricane. Find the strongest room in the house. One with brick walls rather than timber. Throw anything in the garden into the pool. When it gets really rough, climb into the bath and cover it with something solid like a door. Great if you happen to have a spare door in the house.

A police cruiser swung by and, with a horn, advised us to evacuate to one of the designated shelters. If not, try to leave details of your next of kin somewhere safe.

We prepared the bathroom.

Every now and then, out of mischief, I would say, "They won't phone."

And then, suddenly, the danger passed. George had veered away from us and was cutting north to devastate the Panhandle. I was out in the sun by the pool, barbecuing with my White Owl cigar.

Happy days.

· · · · · · · · · ·

We were in love with our house and Spring Hill. We were visiting local realters and viewing homes. A large corner plot down the street from us was for sale, with a three-bed, two-bath and double garage. Fully furnished, it was less than fifty thousand pounds. The owner was relocating to another state and was leaving the Cadillac in the garage.

We worked out the cost of living and made plans and had dreams. However, in the real world, that's all they were. Dreams. The last two weeks were possibly the best of our lives. When it came to go home, we were very, very sad. Barbara later confessed that it was the only place in the world she had cried over when leaving.

We just did not want to leave. On the plane home, I said, "They won't phone!"

Forty-four

We were very happy at Dad's. But!

We wanted a place of our own. A house came up for sale in Valley Road. It was the next road away from the Old Man's. It was virtually at the end of his back garden. It was a sound, solid house that needed a bit of modernisation.

An old person had died in it and the son was anxious to sell. Quickly!

I thought, "I'm your man."

When I found out how old the son was, I realised why he was in such a hurry to sell. He was nearly eighty. God knows how old his mom was.

He wanted to go on a cruise. I thought to myself, "Better keep it short. The Med. A world cruise might be stretching his luck a bit."

The agent wanted fifty-five thousand. I said I could only raise forty-seven. They refused the offer. I went straight to the seller and told him he could have his money as fast as his solicitor could do the paperwork.

Valley Road was mine.

It was a great location. I could still watch over the Old Man, and also Mom was not far away either. We could keep an eye on both of them and still have our own place.

I was happy with my retirement. Everybody kept telling me I was too young for it and how was I going to fill my days? Surely I must be bored?

The fact is, I wasn't bored at all. I'd started writing songs again and found myself continuously jotting down lyrics on beermats, napkins and even tablecloths. Once, in the middle of the night, I grabbed a pen from beside the bed and began scribbling on the bedsheet.

I got a right bollocking and learned from it to keep a small notebook on me all the time. In fact, I kept them strategically placed all over the place. By the bed, on my chair, in the car, in the bar and everywhere I could possibly be.

I soon learned that, when you get an idea, it is crucial to write it down immediately. You can forget in minutes. The next day, it is completely gone for ever.

I became a member of the International Guild of Songwriters and Composers and formed collaborations with various other aspiring songwriters. It was great fun.

One particular composer, Nick, lived not far away from me in Birmingham. He had a room in his house that he had turned into a studio. We used to make demo discs and send them to various record labels, without great success.

We recorded some good songs that the Guild would give an assessment, most of which were very favourable. However, success in music is a tough, hard game. The harsh reality is that, unless you can write, produce and perform your material, you have little chance of even getting anyone to listen to it.

You can imagine my disappointment when Nick dumped me as his lyricist. "Des," he told me, "Are you going through some great emotional crisis in your life? Your lyrics depress me.

They are too deep and meaningful. They upset me sometimes and even make me cry. I don't want to work with you any more."

I hadn't felt this way since the Beatles broke up. First it was Lennon and McCartney. Now it was Des and Nick. I feared the worst for the popular music industry. To be fair, I thought that stirring people's emotions, good or bad, was what good songs were all about.

Well, McCartney soldiered on, so must I.

I wrote a song *"Heart full of Pain"*, inspired by my break-up with Teresa. I submitted it for assessment to the Songwriters' Guild. They loved it and suggested that I get a professional demo CD recorded and bombard the record companies and local radio stations until someone took some notice.

I booked some studio time in London. Howard, the producer, had listened to me sing the song on tape to get the feel of it in advance of my arrival. I wanted a professional singer for the recording, to get the best results. Howard asked me why I didn't want to sing it myself. Because I can't sing, I told him. He then surprised me by saying, "Well, Des, I think you can sing a lot better that a lot of people who think they can."

Not convinced, we used Howard's session singer and, with Howard on guitar and synthesiser, recorded a really presentable song.

I sent it to local radio stations, BRMB being one. At the time there was a great morning presenter, Graham Mac. I was at home one evening when the phone rang. It was Graham. "Loved your song, Des. It's great to know there is still someone out there writing proper songs and not the crap I have to play all day. *I'm Just a Teenage Dirt Bag Baby* for one. If it was my song I'd flood it to every radio station in the country. When I started out, I had to send off over two hundred demo's before I got my first show in Australia."

We went on discussing ideas and our lives and, after a while, he said, "I know where you should send it. Daniel O'Donnell. It's right up his street."

I took his advice and started some research. He recorded for Ritz Records in Dublin. His producer was John Ryan. I phoned him and he said to send the recording. I did and, over the next few weeks, talked regularly. He told me that at the moment Daniel was not recording new songs, but covers of old standards and classics. However, he said, as soon as there was a recording of fresh material in the pipeline, he would call me straight away.

Working on the theory that you can't get too much publicity, I phoned the *Birmingham Evening Mail* and told them the marvellous news. So excited were they for me that they sent a photographer and reporter, Sophie Blakemore, for an exclusive interview.

With a brand-new haircut, especially for the photo shoot, I posed, CD in hand ready to enter the Hall of Fame. Sophie's excitement at her first interview with an international songwriter inspired her to colourful prose that far from did me justice.

The ensuing fame from my press interview was short-lived and I soon faded back into obscurity, along with Peter Sarstedt and that geezer who wrote *Rosy*.

Is he still husking?

I decided to take a break from song-writing and get down to writing my eagerly-awaited first novel.

Alert the media.

My first grandson was born and, along with him, the hero for my book - Jack Harry Reec.

Billed as an anti-hero and compared by some critics to James Bond, I was satisfied to be even remotely associated with Ian Fleming.

"Following in Fleming's Footsteps," was the heading in the *Jersey Evening Post*. I could live with that.

Writing is hard work and no easy task. It can at times be exhilarating and, at other times, despairing. But always exciting and challenging.

Finally, after nearly two years, I had a completed manuscript that I was satisfied with. My mom's sister, Auntie Peg, had given me an Ivory Buddha that I'd had since I was a young lad of ten. She told me that if I light two candles, one on each side of it, and rubbed its belly, I could make a wish.

Why not? I thought, I've got nothing to lose. In the privacy of my study I performed the ritual and wished for my manuscript to be accepted.

I posted it off to Athena Press in London. I waited a long week for a reply.

It read as follows:

ATHENA PRESS READERS REPORT

Private Execution is a rapidly paced and successful exercise in the establishment and maintaining of a powerful level of critical tension. It is a high octane novel of fast living, killers, the IRA, yardie drug lords, ETA, yachts, private planes, bullion robbery, good looking and not too particular ladies, travel, plenty of booze, glamorous locations and big time international shenanigans; it is a powerfully effective and entertaining mix.

In particular, Reec is a top class good bad guy, and he is really bad. There is a whole cast of other bad guys, most of whom end up being shot, blown up or in one case hanged by "our hero". I read this novel on an uncharacteristically beautiful afternoon in my Oxfordshire conservatory, and it was a great companion for a glass of chilled Chardonnay. I hope Reec comes back in another adventure. He's too good a character to use up in one book. It is still rather a novel idea to have a hero who is sympathetic but also entirely amoral.

This is a lively and readable pacey action novel, the narrative moves along at a great rate, dialogue is particularly well crafted, the characters are all believable, and in particular the denouement is strong and effective. It is written with drive and conviction. It is definitely a "boy's book";

I cannot see a great female readership here. The diving background is particularly well detailed and lends an underpinning of reality to this highly improbable top level thriller.

We will be pleased to offer Desmond McGrath a publication contract for PRIVATE EXECUTION.

Signed Mark Sykes

Consultant Editor-in-Chief

It sounded to me that he liked it.

The next six months were to be some of the most exciting of my life. The whole editorial process was long. There were contracts to sign, editing to do, proof-reading and the final layout.

Then came the cover design. I had a strong opinion of what I wanted and submitted my own art ideas. I wanted a diver swimming towards the wreck of a sunken Lear Jet, with gold bullion bars spilling from the hold onto the sea bed.

The team did me proud. It was just what I wanted. Then there was the cover blurb.

A plane carrying gold ingots crashes into the sea. A boat with its illegal cargo of drugs flounders on the rocks. These two events shape the future of our tough modern-day hero, Jack Harry Reec, whose skills as a diver and willingness to take risks bring him big profits.

The action takes him from Jersey, his native island, to Spain where he lives life to the full, enjoying good food, excellent wine and the beautiful scenery with his new girlfriend and his long-time male friend. But the relaxed and self-indulgent lifestyle does not last long. Along the dangerous road he has followed, he has made enemies of violent, brutal men with horrifying and bloody results.

An ex-Falkland's veteran, Jack is already haunted by death. Now he is ruthless in his pursuit of vengeance for the latest tragedies. We follow his quest to "Rid his soul of anger". Carefully and patiently, he plots a terrible revenge on his enemy.

But the final stage of his plan may be jeopardised. Will there be a *"Private Execution"*?

This at times shocking action drama is a genuinely exciting story, full of surprise and tension right until the end.

When I received the proof copies, I was the happiest man in the world.

I had achieved my life's ambition, and the best was still to come.

I lit more candles around my lucky Buddha.

Forty-five

1999. We had been in Valley Road for a couple of years. It was nicely decorated and well-presented, but we had resisted the temptation to spend a lot of money on it. If we knew we were staying we would have had new windows, kitchen and bathroom.

However, we had no intention of staying there too long. As it stood, there was a good profit in it without the expense and upheaval of complete modernisation.

It's a good job I was retired, really, as watching over two elderly parents in two separate houses was quite time-consuming. Dad went to his local pub every day at lunchtime. It was an easy walk and it gave him exercise. Two or three days a week, I would pop in and have a beer with him.

Mom was fiercely independent and well able to shop for herself and do all her own cooking and housework. She only really needed visiting.

One Friday evening, I had a call from the Old Man. He said he didn't feel well. I went around with Barbara to check it out. He didn't seem too bad to me. I just thought he was having an off day. After all, he was eighty. He'd been to the pub and was having a whiskey and a cigar.

However, Barbara decided to call an ambulance. She is the family medical expert and saw things, quite frankly, that I didn't. I thought she was wasting the emergency services' valuable time.

Of course, I was proved to be wrong. The ambulance men did some tests and decided that he needed to go to hospital. I still couldn't see the point, as he didn't really look ill to me.

Wrong again. They kept him in. He was in all weekend and going crazy to get out and go to the pub. He asked me to bring him in some Scotch. Obviously, I couldn't do that. He was taking all sorts of medication. It might have killed him.

Myself and Barbara were visiting him as usual and were just a few yards from the ward he was in when pandemonium broke out. Warning sirens were going off and doctors and nurses were charging all around the place with trolleys and carts. It was a crash team. The Old Man had gone into cardiac arrest.

The next few hours were frantic. They brought him back and put him in Intensive Care. He never came out alive. We gave him a fine funeral and he was buried with my two brothers, Philip and Edward.

Dad had been organised with his affairs and, after leaving Mom adequately taken care of, he left me and Barbara his house and shop on the Coventry Road. He was the sort to always discuss things, and it was his dear wish that Barbara and I should move into his house after he was gone.

I had no problem with that and put our house in Valley Road up for sale. It didn't take long to sell and we moved into Dad's just before Christmas. The next six months were spent modernising the house and, when it was done, we decided to have a holiday.

I loved Las Vegas. Barbara had never been there and was a little dubious as to what it would be like. It certainly wasn't Florida.

Barbara's older sister, Jane, had never really been anywhere and, for some reason or another, she got in on the holiday. I didn't mind at all. Jane was all right and I got on well with her.

As I already knew the score, I decided that I would organise it all myself. No point wasting money on travel agents when you can do it yourself. I got the cheapest flights available with Travel Bag. I then got details of various hotels from a pile of travel catalogues and decided on a hotel I fancied.

The *Gold Coast Hotel* was just off The Strip. It had all the usual amenities: casino, bars, shops, restaurants. The phone number was not in the brochure, so I got it from Directory Enquiries. I was amazed at how cheap the rooms were. They were all twins, with two double-size beds in each room. We definitely only wanted one room. The cost of two rooms was prohibitive, and out of the question.

The only reservation I had was Jane in the same room. It was a ten-day vacation and I was concerned whether, with her reputation as a man-eater, she would be able to keep herself under control.

It was a well-founded concern. On more than one occasion I was to wake and find her next to me in bed! Or was it me next to her in her bed?

Anyway, no worries. Happy days.

It all made for a bit of amusement when I was making my speech at mine and Barbara's wedding a few years later.

Funny.

The trip turned out to be a great success. To my surprise, Barbara and Jane both loved the place. The shopping was great and they even bought Barbara's wedding dress for half the price of the exact same dress in England.

I had one amazing run on the roulette table. I was playing for a straight eight hours, with only a toilet break. I was playing a low-stake table and, at one stage, was up many times what I started with.

The waitresses were continuously serving complimentary drinks and snacks. I had fourteen Mai Tais and inevitably started to gamble a little erratically. After eight hours, I was back with what I started with. I cashed in my chips.

I'd had eight hours' great fun, fourteen Mai Tais, food and it hadn't cost me a penny.

There was a swimming pool and bar where you could top up your tan. We spent a few afternoons there.

We decided to have a few days in San Diego for a change. We started out early one day and drove the whole way with only a couple of breaks. However, it was not as good as I had remembered it from my trip with Ed and, after three days, we drove back to Vegas to see out our holiday.

• • • • • • • • • •

At about this time I was told about a specialist butcher in a small village called Knowle, Eric Lyons, who was looking for some part-time help. I called in one afternoon to see what was involved. I don't like to admit it, but I was missing the banter in the trade and the general fun of the game.

I got on with the owners and it was agreed that I start for a couple of mornings a week and see if we both got on and whether it would work out.

With the owners there were six staff, all full-time. Also, there were a lot of various part-timers and young lads to do the cleaning and unskilled jobs.

They made all their own pies, sausage and cooked meat and had a well-organised, thriving business. From the beginning they made it plain that I would be a great asset to the company. My butchery skills were legendary in the trade and, with my years of expertise and knowledge, I could only see me taking them from strength to strength in every direction.

I was enjoying it immensely, which was obvious when I received my wages. I wasn't there for the money. Mom loved her pies and, as part of the deal, she could have as many as she could eat. Also, they cooked a great breakfast and you could have a sausage roll if ever you were hungry.

I signed up and did nearly fifteen years, receiving a T-bone steak as my leaving present.

Seriously though, I really enjoyed it. The one thing I didn't want to do was serve in the shop. I was fed up with that part of the job and confined myself to preparing the meat for sale in the back. I also took care of the deliveries and made sure that nobody was taking advantage by giving us short weight, a practice that was rife in the trade.

One particular wholesaler took offence at my checking but, unfortunately for him, I explained that if I was signing for something then I was going to make damn sure that it was all there. It was my head on the block if it wasn't. It was all too easy to run into a shop and dump a load of meat, then stamp your feet for a signature because you were in a hurry. It was amazing how many times there was an item missing, that miraculously appeared when asked where it was.

I was just nearing completion of my book when I started, and everyone was as excited as me when it got published. Of course, everyone wanted to read it, but nobody wanted to buy it! Having said that, it created a lot of banter and it was nice to share my own excitement with my friends. I would have missed out on all that if I'd been stuck at home. Also, as I was still going to the market every week, I was having my leg pulled there, too.

When I revealed that I was starting a second book, there was no end of suggestions from the lads at the shop. One in particular, Dereck, had a very vivid imagination and was always coming up with lots of suggestions. Most were pretty crap but, by the law of averages, he did bounce some useful ideas off me, which I found stimulating if nothing else. As I told him once, "You ain't as dumb as you look," to which one of the other lads chorused, "He couldn't be".

I once read an interview by somebody famous, who said, "Success is ninety per cent luck and ten per cent talent."

I'm not too sure if I fully agree with that but, in my case, it was certainly true to an extent.

When it was published in Great Britain it was simultaneously released by Athena's parent company in the US.

It was on Amazon and all major book sites. I wanted to see what was being said about it on the net. When I looked on Amazon, I saw an image of the book and all the info about it.

But, better than that, directly above it was the headline: *Harry Potter and the Half-Blood Prince*, £9.99 41% off.

So what? you might think.

Well, it meant that, as it looked, my book was the Harry Potter book. Anybody looking for Harry Potter saw my book. With millions of people worldwide going mad for Harry Potter, I was getting massive, unexpected exposure. It also said,

"Only one left in stock - order soon (more on the way) $8.46." Someone commented, "Looks very exciting. Must buy one."

There was interest from the local papers and I was enjoying my fifteen minutes of fame. I soon learned that it's true what it says in the Bible, "A man is never a prophet in his own town".

I had to settle for my critical acclaim and encouragement, not from those closest to me, but from my peers and fellow academics. I'd heard from a friend how a certain family member had declared it a load of crap and threw it down only half-read. There was too much violence, sex and bad language in it for his liking!

You can't please everyone, that's obvious. I took it on the chin. Everyone is entitled to their own opinion. If we all liked the same thing there would be only a handful of writers in the world selling books.

I didn't let it spoil the fun, though. I was already planning a second Jack Reec novel.

However, to avoid negative vibes, I kept it low-key and told only close friends and the staff at the shop.

No negative vibes there.

By the way, I have to say, my mom loved it and wallowed proudly in the glory of her son being an author. She kept all my press cuttings and, even at her ripe old age, was not offended by the sex, violence and bad language. She did, however, question the language, saying it was a bit excessive.

I explained to her that I had aimed the book at the American market and that was what they wanted in the US. As Graham Young from the *Birmingham Evening Mail* once said to me,

"Des, if you're going to be a success, America's the place to do it."

In the middle of all this excitement I received a posh-looking letter in the post. It was from Spain.

It was a wedding invitation. It read,

Miss Claire Louise Baker and

Mr Matthew John Francis request the pleasure of Des and Barbara to celebrate their marriage at TIKITANO BEACH RESTAURANT

CTRA CADIZ, 164 ESTEPONA, SPAIN

on Saturday 3rd June 2006 at 4.00pm

And afterwards for a champagne reception followed by dinner and dancing till late.

Claire had been a close family friend since birth. Her mom and dad were great friends who we socialised a lot with and attended each other's barbecues. Matt had been seeing Claire for a long time and everyone was very happy for them both.

Matt's mom and dad, Trevor and Helen, now lived in Marbella. Trevor was a retired footballer who had once played for Birmingham City and Nottingham Forest. Helen was a loving wife and mother and possibly one of the nicest ladies I have ever met in my life. She was a beautiful, thoughtful, down-to-earth person who had been totally unaffected by the fame and attention thrust upon her. They were, together, probably the Posh and Becks of their generation.

Barbara and I were excited and looking forward to the wedding. I phoned Matt and told him I wanted to stay ten days. Part of his business interests was property maintenance and lettings. He said he could also arrange cheap flights and car hire. I gave him our dates and he arranged everything.

Return flights from Coventry with a hire car waiting at the airport. The accommodation was a two-bed apartment on top of theirs on the gated golfing community where they lived. Helen's mom and dad would be stopping in the apartment above ours. It had all the ingredients for a lot of fun.

Matt and Claire's apartment was on the ground floor, with access to the gardens of the whole complex. There was a huge, communal swimming pool and a bar and restaurant on the golf course. Matt told me that other friends and family had also rented places on the complex and it would be a great community atmosphere.

It sounded great and we couldn't wait.

I didn't realise how many Blues fans there were at the shop, until I casually told them I would be away for two weeks as I was going to a wedding in Spain.

As the only Villa fan in the shop, I took a lot of stick. No worries.

Happy days.

Forty-six

The drive from Malaga airport was a nightmare. How we got to the apartment alive was a miracle. The standard of driving in Spain is horrendous and, apparently, the road from Malaga down through the Costas is the most dangerous in Spain. We needed to get to San Pedro. With Barbara actually in tears, we pulled off the road and made a call.

Matt answered. "Where are you?" We told him exactly.

"You're practically on the doorstep. I'll guide you in."

With me on the phone Barbara followed directions and, in ten minutes, we were at the security gate to the complex. Matt had cleared our entry.

"I'll have a drink ready when you get here," he told us. "Thanks, we'll need it."

We drove down a long road that took us to the community. Matt was waiting for us and directed us to our designated parking spot in the underground garage. Leaving all our luggage in the car, we followed him in the lift up to his apartment. Claire was waiting with a bottle of cava. Greedily we set upon it.

"Christ," I said, "That road's a bloody nightmare." "You get used to it," laughed Matt.

After catching up on all the gossip, Matt and Claire showed us up to our apartment. It was luxurious and laid out exactly like theirs.

"It's never let out normally," Matt informed us. "The owners keep it exclusively for themselves. But when I told them you were personal friends of mine they said you could have it for a couple of hundred pounds for the two weeks. Is that all right?"

"Of course," I told him. "It's an absolute steal at that."

"Just go easy on the air con," said Matt. "Only have it on when you're here and don't leave any doors open when it's on."

"There's no food," said Claire, "So I've got a meal cooking for you for tonight. Mom and Dad are already here, so I thought it would be nice for all of us to have the evening together."

"Great," we said. "See you later."

Ivor and Elaine, Claire's mom and dad, met us at Matt's for dinner. We had a nice evening and they agreed to meet us in the morning to take us to the shops and fill the car with food and drink. It was a bit isolated on the golf community, so you didn't want to be going shopping every day. As with everywhere, there was a local shop that was obviously a lot dearer, but a godsend when you ran short of something. We spent the rest of the day by the pool topping up the tan.

Later in the evening, Trevor and Helen arrived with her mom and dad, who they introduced and installed in the apartment above us. They were a nice old couple and I got on instantly with her dad. We both shared the same common things. We both liked our beer and, spookily, the same cigars, White Owl.

There were a few days to go before the stag night and wedding. Matt organised a big barbecue in the grounds outside his balcony. Lots of the wedding guests who had arrived early for the wedding were there. It was very busy. The barbecue was smoking with all sorts of food and there was a shedload of beer and wine.

Helen knew I got on well with her dad. She said, "Des, do you mind keeping an eye on him? He has a habit of wandering off."

"No problem," I reassured her.

It was a typical hot Costa afternoon. Helen's dad was feeling it and complained of being tired and hot. I suggested we got a few bottles and found somewhere quiet in the shade.

With a couple of beers apiece, we wandered off along the grass to the shady side of the building.

"This looks OK," I said.

We sat down with our backs resting on the wall next to someone's balcony.

"Fancy a cigar, Des?" Dad asked, producing two huge cigars. They were White Owl New Yorkers. I didn't know they did them that big.

"Got them from Gibraltar, Des," Dad said, in his broad Welsh accent. "Bloody lovely they are."

They were, too! With our beers and cigars, we sat swapping tales until a very refined, female voice said, "Enjoying ourselves, are we?"

We looked up at a very smart, elderly lady leaning on her balcony taking deep whiffs of our cigars.

"I smelled your cigars," she said.

We both instantly started to apologise.

"No, no, don't apologise," she was inhaling deeply. "I love the smell of cigars."

We were passing the time of day when the search party arrived. They were sent out to look for us.

"I'm all right, I'm all right," Dad protested. "I'm with Des."

"Yes, we know. It's Des they're worried about."

• • • • • • • • • •

It was the night of the hens and stags.

The hens were going to a restaurant on the golf course. The stags were going to a restaurant in Puerto Banus, then some lap-dancing bar. There was a coach to collect us all from various pick-up points.

We were first pick-up at the golf complex. There was Claire's dad, Ivor, and his son,

Matt, myself and some more men that I had met earlier. The next stop was a building-site in the hills. A singer friend of Trevor's, Paul Carrack, had just bought one of the few finished villas on the site. I hadn't heard of him myself, but recognised some of his songs and the bands he had played in. The obvious song was *"In the Living Years"* by Mike and the Mechanics. The other band was Squeeze. He didn't look like a rock star. Short and bald.

Another person I recognised was Barry Fry, a football manager. He had an even bigger cigar than mine permanently jammed in his face. There were many footballers and managers, the most notable Gareth Barry, whose stag do in Las Vegas I was later to be invited to.

The restaurant had been totally reserved for the night and was full. It was all very well organised, with tables set and labelled for different numbers of people. I found myself with three others. Two I'd never met before and one who I knew from the apartments. He was staying close by me. I can only remember the one name, Dennis, because we got on particularly well.

I'd never imagined myself being in the company of so many celebrities and millionaires.

Strangely, though, I never felt intimidated and mixed in easily. Myself and my three table companions introduced ourselves and got stuck into the bottle of champagne that was chilled in the middle of the table. We conversed easily and ordered our food. We all had fillet steak. It was delicious.

"Matt tells me you've written a book, Des," said Dennis. "What's it about?"

I briefly ran through the plot and all seemed genuinely interested. We each divulged some interesting snippets about each other. We were having a lovely evening. Matt came over and asked if we were all right. We were.

"What did you think of the fillet steak, Des?" Matt asked.

I told him it was great. He explained to the others that in another life I had been a butcher. They knew.

Matt seemed pleased and moved on.

I don't know how, but the conversation got around to shoes.

"I get all my shoes from Italy," divulged Dennis.

"I get mine from Aldi," I confessed, which was true; I'd recently bought two pairs of slip-on moccasins for about ten quid a pair.

Dennis said something polite and I thought no more about it until I was talking to Matt the next day. I mentioned the shoe conversation.

Matt started laughing. "So you don't know who he is?"

I shrugged. "Dennis."

"Have you heard of Dolcis?" Matt asked. "They have stores in every city in Britain.

Dennis owns them."

"Oh," I said, "I hope he didn't think I was taking the piss when I said I got mine from Aldi."

The stags moved on to the bars in the centre of the port. Arseholes on billion-dollar boats were prancing the decks with half-naked Wags. Prats in Ferraris were doing wheelies around the dock and wankers without a pot to piss in were posing like decorations outside the heaving bars, where idiots were paying a month's wages for a bottle of cheap, fizzy plonk.

I hated the place.

Dennis and his companion had been lost in the throng. I was with the other guy from the table, who lived by me. Excitedly, the word was going around that the stags were all going to a lap-dancing club. Entry was free to us.

They told us where it was. We said we would see them in there.

I said to my table mate, "I don't know about you, but there's no way I'm paying ten quid for another bottle of this warm piss."

"I'm glad you said that," he replied, "And I don't fancy that seedy lap-dancing club, either."

"I must be getting some sense at last," I told him back, "Because I don't like strip joints, never have. Fancy getting a taxi home?"

He was all for it, so we found a taxi rank and I asked "how much?" I never get in a foreign taxi without asking first.

"Twenty Euros, senor."

Ten each. That would do nicely.

• • • • • • • • • •

The wedding had finally arrived.

A fleet of stretch limos slowed to a stop outside the apartments. Climbing into one, we were chauffeured to the *Restaurant Tikitano* on the beach. Greeted in style with champagne, we mingled outside in the spacious grounds where the ceremony was to be held. As the time approached, we were guided to an area overlooking the ocean that had been set out with rows of seating.

The bride and groom arrived and the wedding was held in glorious sunshine. It was late in the afternoon and it was not too hot, just pleasantly comfortable. When all the rituals were concluded, it was back to the gardens outside the restaurant for more drinks and mingling.

At the given signal we were all to take our allotted seats. I found our places and was pleasantly surprised that Matt had sat us with Paul Carrack and his wife. He thought we would be comfortable with each other. Barbara had made friends, at the hen do, with his wife and we were quite happy to share their company. Barbara chatted away with Paul's wife. Paul and I talked about music, and shared interests. He had many fascinating stories to tell about his career and I told him how it had taken me twenty years to become an "overnight success" with my writing.

"I know the feeling, Des," he said. "We've all been there."

I asked if he was going to do a turn later. He told me that nothing had been arranged, but he felt sure that at some point in the evening, after a few more drinks, he would be dragged up for a few numbers.

I'm not sure if he was being completely truthful as, later in the evening, accompanied by one of the most popular DJs in Marbella, he did an amazing act.

Sweating profusely, he sat back down afterwards and had a few drinks. A few people came over to congratulate him on his show. Toni Iomi from Sabbath joined us for a chat and I was tempted to ask him if he knew what had become of Teresa, Bill Ward's ex-wife.

I decided against.

Dennis, my friend with the Italian shoes, wandered over.

"Can I join you for a bit, Des?" he asked. "I'm sat with the biggest arsehole in the place and he's doing my head in."

"Sure," I said. "I'll talk to anyone."

The arsehole he was referring to was the chairman of Crystal Palace football club. I can't remember his name but knew of him and his reputation.

He was a young, fit, blond, handsome, immaculately dressed multi-millionaire with a private jet, who continuously had beautiful women hanging from his arm.

I hated him instantly.

I must have had enough to drink, because I found myself actually dancing at some point in the evening. Eventually, Barbara and I decided it was time to call it a night. We said our goodbyes to everyone and Paul said that if he was ever performing in Birmingham, to say hello. We found a taxi and went home.

The next day we saw Matt and Claire off on honeymoon and, the day after that, it was time for us to face the nightmare journey to Malaga Airport. I have to give credit where it's due, to Barbara. She got us there without incident. I had offered to drive but, as much as she had been dreading it, she still felt safer driving herself.

She's not a good passenger. I am.

I was looking forward to getting home. With good reason. A couple of weeks before we left I had received my first-ever royalty cheque. Myself and Barbara had gone to Mercedes Benz Direct in Balsall Common and purchased an SLK200 sports. While I was away, they were preparing it and transferring my 52 DBM number to it.

We took a taxi from the airport straight there and took delivery.

It was a beautiful day to drive home in my dream car.

I still have it.

Forty-seven

On Monday morning it was back to work at the butchers. I loved Monday, it was just a tidy-up day and that was my speciality. I was to go through all the fridges, sorting out all the bits that hadn't sold at the weekend and preparing them for sausages or ready meals.

At ten we all had a break and gathered by the office for tea and Cornish pasties. It was there that we shared our gossip from the weekend.

Everyone wanted to know how things had gone at the wedding. I gave a quick low-down. The only thing the lads wanted to know was what footballers were there and what they were like.

"They were like blokes on a stag do and a wedding. What do you think?" I said. "They aren't any different from us. It's just a job. The only difference is that they're redundant at 35 and have to join the real world a little later than the rest of us."

"An interesting spin," they thought.

I told them that Matt had got me an invitation to Gareth Barry's stag do in Las Vegas and that Gareth was paying for everyone.

Breakfast was soon over and we worked then until 1.00 o'clock. That was dinner-time. Dave the gaffer always cooked a huge tray of bits, chops, sos, steak, chicken - the lot. The lads took turns at buying a loaf, which I always went for. It gave me a chance to nip into a charity shop for a bargain.

I was beginning my second book, *"The Executioner"*, and part of the plot involved the assassination of the Manchester United captain on the pitch during an evening game.

I was deciding at the time whether I would use Birmingham City or Aston Villa. It seemed that I was the only Villa supporter at the shop, apart from Dereck. I was persuaded to use Birmingham. At the time, both clubs were in the Premiership, which they had to be to be, playing Manchester United.

I made a call to the City ground and spoke to various people, explaining that I needed to poke around behind the scenes to find a location to take out a footballer on the pitch and get away without being caught.

They said they would get back to me.

I was watching TV in the evening with a can of lager, as was my habit, when the phone rang. I never bother to answer the phone as it's never for me. Barbara picked up.

"Yes, he's here, I'll pass you over," she said to the phone. "It's Karren Brady for you."

I took the handset. "Hi Karren, how are you?" Pause. "Yes, I'm good too. What can I do for you?"

We chatted for a few minutes and I was frantically signalling Barbara for a pen.

"Sure, that would be great. I'll just jot down her number. Yeah, I really appreciate that, thanks for ringing, Karren, see you later. Bye."

"What are you up to now?" demanded Barbara. "Nothing," I protested. "That was Karren Brady."

"I know who it was. Why was she phoning you?"

"She thinks it's a great idea using Birmingham City in my book and wants to show me around the ground. I suppose she'll probably expect me to take her for lunch but, well, you know how it is. Sacrifices, sacrifices. The things you have to do."

As it happened, I didn't have to take her to lunch. Instead, a really nice girl met me at the ground one morning. "Karren's told me all about you," she said. "You're to have the complete run of the ground."

A groundsman, or somebody, was assigned to keep me company and answer any questions I might have. He was extremely helpful and knowledgeable. I think it was the dressing rooms where I saw a small brick that was engraved,

"Trevor Francis forever a Blue Nose."

I couldn't help myself. "But didn't you sack him?" "Don't want to talk about that," he said defensively.

"Just kidding," I laughed. "He's a friend of mine. I've just been to his son's wedding in Spain."

We were still friends. He asked me, so I told him all about it.

Notebook and pen in hand, I thanked everybody for their cooperation and promised to send them a copy of the book.

Unfortunately, it was not going well. Writer's block. I'm very critical of myself and realised that I was not happy with what I was writing. I reasoned to myself that, if I didn't like it, how could I expect anyone else to like it?

I struggled along and scrapped a lot of what I'd done. It all seemed hopeless. I took a break. I tried again. I took another break. Months had slipped into years.

The Blues got relegated. Another setback.

One day I was talking to Dereck at the shop. We were always trading ideas and banter about the book. Then something was said. I don't know what it was, but it got my imagination back to work.

The first thing I needed to do was rewrite the "hit" at Villa Park. It had to be changed to there, as the Blues were now in the Championship. John Lyons, one of the shop owners who did the baking in the kitchens, knew someone at Villa Park.

He arranged for us to have the freedom of the ground. His name was Pete Dawes. John and I met him one Wednesday afternoon at the *Bird in the Hand* pub on the Stratford Road, Henley in Arden. We climbed into Peter's Range Rover and he took us to Villa Park.

Entering through the staff area we went to the Stadium Manager's office and met Tony, the Stadium Manager. He was a great bloke and very enthusiastic to get involved. He made us coffee and broke out the biscuits, "Reserved for special guests".

He reckoned he knew the perfect location for the assassination and how to get there and escape in what would be pandemonium and a lock-down of the ground. He took us up through all the nooks and crannies of the ground until we were on a gantry behind a block of floodlights that overlooked the goal at the Holt End.

The executioner had a perfect sight of goal and a rest for his rifle. The plot was that Danny McAlinden, the Man U Captain, was taking a penalty at the Holt End. As he kicked the ball Liam Dooley, the assassin, would shoot him in the back. Danny would crash to the pitch and, having scored, the crowd would go mad.

Before people even realised what had happened, Dooley was beginning his escape using Tony's carefully worked-out plan.

My adrenalin was in a rush and it all started coming together. After months of hard work, I finally had a completed manuscript.

However, now came the real test of my endeavours.

I had to get past the publisher's reader.

I lit my candles and stroked the Buddha's belly. Then waited for the Reader's Report and possible rejection of all my efforts, with a simple rejection slip.

It was a long two weeks.

ATHENA PRESS READERS REPORT

AUTHOR TITLE DATE

DESMOND MCGRATH THE EXECUTIONER 12TH DECEMBER 2008

This new novel from Desmond McGrath works well. Like its predecessor, it is what Graham Green once called "an entertainment". It's not a literary novel; it makes no pretensions to changing the face of the cultural history of our century. It sets out to shock, surprise and entertain, and in this it succeeds.

Both Liam and Daniel are very accessible characters, and it is not hard to identify with their adventures. THE EXECUTIONER is a strong, gritty, narrative driven page-turner of a read, and these events are not, in fact, removed from real life. The novelist has been able successfully to create, and people, his landscape. The novel is driven, yes, eventually by plot and character, but also by a sure hand with dialogue and with place, which effectively advances the narrative. The whole idea makes convincing sense within the boundaries created by the author. The scenario on which the plot is based is quite possible albeit with a stretch of the imagination.

All of the secondary characters are well and convincingly characterized.

McGrath successfully creates in the reader the suspension of disbelief, which has been the task of the creator of fiction since storytelling began, and gives him or her a few hours of escapist pleasure. This is a pretty good result. I am sure I am not the only person who may appreciate this gripping and well-structured novel that has a consistent texture and, within the fiction, a strong reality. Without reading too much into it, it can perhaps be taken as a metaphor for the world around us.

We will be pleased to offer a publication proposal for this gripping new novel, and if we go ahead with the project, we will, as is now normal procedure for us, publish both in the USA and the UK.

Mark Sykes Editor-In-chief Athena Press

I didn't think it could get more positive than that, so I decided to let him have it. Happy days.

Forty-eight

Excitedly, I showed the letter to Barbara. She read it and handed it back to me. "That's great. Congratulations," she said, with a little less enthusiasm than I was hoping for. I suppose really that nobody is going to be quite as excited as me and perhaps I was expecting too much.

I was at work the next morning and first thing I did was show the letter to Dereck. He, by comparison, seemed even more excited than me. John in the kitchen, who had come to Villa Park with me, wanted it next. Stan, the shop manager, was delighted. He was one of my biggest fans and had loved the first book so much that he was promoting it to his neighbours and selling copies.

Gill and Vida, the girls in the kitchen, were also fans and it was looking like there wasn't going to be a lot of work done that day. There was a rumour going around that Pete and Dave, who owned the business, were thinking of declaring a national holiday.

"But without pay!"

That got them working again!

The next few months followed the same path as the first publication and soon it was time to unveil my idea for the cover and the all-important blurb. I'd told the art people exactly what I wanted, and it was all I could wish for and even better. They also told me that, as they were re-launching the first book, they thought it would be a good idea to revise and improve the text layout for book one. "We feel these changes will help the visual identity of the books as a series."

I was chuffed with the results. The cover blurb read as follows:

The executioner's face is always well hidden – isn't it? When Liam "The Slaughterman" Dooley, ex-IRA hit man, descends on Jack Reec's Andalusian farm to avenge five gangsters Jack has disposed of, he doesn't bother to hide his face or his intentions. But when guests are kidnapped and Jack's stunning girlfriend, Barbara, goes off on a blind date and never comes back, sympathies change and Jack is joined by the Irish brigade and his Spanish friends in a manhunt for a savage killer.

David Walker is the quarry. Smooth, fit, handsome. A man with a mission -bedding beautiful women. And a man with a beast in his head, shouting a deadly message.

The hunt for the psychopath takes Jack and his friends through the hills of southern Spain and across the Straits to the Rif in Morocco. Stunning settings, dramatic showdowns on land and sea, and a duel to the death in North Africa highlight an action-packed, page-turning thriller. Keep reading!

I showed all my friends at work.

"When's the party?" everyone was asking.

What a great idea, I thought. I made some enquiries and eventually booked the Cotswold Room at the local ex-servicemen's club the *Ivy Leaf*. The management were great and organised food and a DJ.

I invited all my friends and alerted all the relatives whom I normally see only at funerals. They all turned up and, with a pile of books to sell and sign, we had a great party. Two particular friends from the Poultry Market made the night, Dean and Judy.

Judy was by far my biggest fan. She was a trolley dolly for one of the airlines out of Birmingham and tended to get very drunk. She had what I can only describe as a very pronounced chest that was barely contained in her less-than-adequate top.

She followed me everywhere like a groupie. My daughter, Angela, thought it was hilarious and took the piss out of me for days afterwards.

Somebody else gave her a great nickname that I thought would make a great character, "Judy Goody Two Boobs".

Hey ho!

I was on fire. My head was bursting with ideas. Every spare moment I had I was writing. Whereas the second book had taken years, within nine months I had finished the third, *ONE LAST KILL*.

I thought it was really good. To me it was the best book yet. I had a trilogy. But, of course, it didn't matter a jot what I thought. It was down to what the publisher thought.

I sent away the completed manuscript and waited apprehensively for a result. It was three weeks and I hadn't heard a thing. Self-doubt began to creep in. Then, one morning, a familiar envelope dropped through the letter box. I opened it. It was the Reader's Report!

ATHENA PRESS READER'S REPORT

Author Desmond McGrath Title   ONE LAST KILL

Date 20th August 2010

One Last Kill is a rapidly paced and successful exercise in the establishment and maintaining of a powerful level of critical tension. It is a high octane novel of personal interaction, powerfully expressed; it is an effective mix. It sits well with its predecessors.

It plugs into topical preoccupations; the imminent threat to the prosperity and lifestyle that most of is in the "West" take for granted posed by organised crime and the advent of the various incarnations of tough egg international criminality, and the rather exciting although equally alarming prospect of the activities of shadowy groups that may or may not threaten to touch us all.

It is literate, it is carefully and originally structured and it is rather what a certain market sector wants today, although it in no way gives any feeling that it has been created as a commercial exercise a la Jeffrey Archer. Some might think it a little anachronistic, a little "old fashioned", a little too comfortable in its skin. But there are many who like to read a book of this proved and tested genre, and our hero - or anti-hero - is a powerful and effective characterization.

McGrath is a competent writer. His method and style has, of course, echoes, but it works. Character, situation and narrative are three dimensional and believable. Our man is a credible protagonist and the baddies are real baddies. The language and attitude are contemporary.

This is a lively and readable novel. McGrath has a vivid and productive imagination and his disturbing scenario is as likely and credible as any, and it rings true. The narrative moves along at a great rate, dialogue is particularly well crafted, the main characters in particular are completely believable, and the denouement is strong and moving. It is written with drive and conviction. McGrath allows himself the space fully to develop this theme but he does not lose his way, although the narrative is complex. Although located firmly in blockbuster territory, this is better defined as a "literary thriller". McGrath is not attempting to rewrite the cultural history of the modern novel; he is giving the reader a few hours of pleasure and a couple of things to think about. And this, after all, has been the aim and objective of the creative writer of fiction since the first storyteller unrolled his mat in the market square.

We will be pleased to offer a publication contract for *ONE LAST*

KILL. Mark Sykes

Editor-In-chief

Athena Press

Phew. What did you make of that? I had to read it three times and get the dictionary out to understand some of the adjectives.

Anyway, we were rockin' and rollin' again. Happy days.

Everything was going great guns and we were nearing publication. Then all correspondence stopped. I couldn't get through on the phone and nobody answered my letters.

I was really getting frustrated with it all. I had a bad feeling that something bad was happening.

I wasn't wrong.

I got a letter from the Official Receivers.

Athena Press had gone into liquidation. All work had stopped and a takeover bid was being looked at.

Disaster.

Forty-nine

I was in limbo. Somewhere out there was my fully-edited manuscript ready to go to print. It seemed I wasn't the only author with problems, there were loads of us. I was told that even if somebody found it, it was the property of the receivers until a buyer for the business was found.

A company was in negotiations - Fast Print Publishing. I made a phone call and eventually got connected to a nice lady, Pauline. She was very friendly and helpful. She said she would do her best to track down my missing manuscript.

I just had to be patient.

To amuse myself, I started plotting ideas for another book, *"A DATE WITH DEATH"*.

It was going to be another Reec adventure set in Florida. Caviar, huge American singing sensation, was going to prepare for another triumphant world tour.

She would need a support band.

John from the shop, who had instigated the Villa Park tour, was always going on about this great band that he had seen at some muddy-field rock festival. He was still a secret closet air-guitar hero. The Quireboys was the name of the band. I told him my idea for the new book and suggested that he seek out their management and offer them the opportunity of worldwide fame by being support band to Caviar on her world tour.

With a bit of spare time on his hands and a good knowledge of the internet, he got to emailing Bruce, their manager.

Bruce was extremely keen on the idea and said he would put it to "the boys".

After a few weeks of emails, it was arranged for us to meet the band at some big club where they were performing. John asked if I was interested in going to the gig. Of course I was. It would mean staying in a hotel overnight. Great. I let John make all the arrangements.

John knew his way around the country and was far better at driving than I was. I never went far. Between us we sorted out the supplies we would need for the journey and overnight stay. I had a camping cool-box with beer, ice, gin and tonic. John brought the snacks for the night. Top of the list was one of his famous homemade pork pies.

We were set.

John drove us to our Travelodge, where we parked up. The plan was to get a taxi from there to the gig. We had some refreshments in the room and called our taxi.

The venue was not what we had expected. It was a bit rough. We found security and told them we were there to meet the band. We were expected. Paul the guitarist and Keith the piano-player greeted us excitedly. They wanted to know all about the book. We wanted to know all about the band.

Paul was not too happy. "Des," he said, "I can't understand why we're playing a shit-hole like this".

We were backstage and looking out over a vast, non-seating area. Keith was equally scathing. "We've played to seventy thousand people, in arenas, Des. I can't understand what Bruce was doing booking us in here."

Not wishing to pour petrol on the fire, I told him how much I was looking forward to the show and talking to Spike, the singer.

The venue was getting full. The bar was crowded. We said that we would see them after the show. We made our way off the stage and pushed through the crowds to try and find a quiet spot. John went to get a couple of pints.

On stage, Paul was pointing to somebody and then me. Bruce, the manager, was looking for me. Tapped on the shoulder, I was led to a grotty area behind the scenes and introduced to Bruce. He was a smart and friendly man in his fifties, dressed in an immaculate suit and tie.

Bruce led me through a couple of decaying rooms that passed as dressing rooms and introduced me to Spike, the lead singer. He was wearing his trademark red bandana and was having a beer before the show. I told him of his part. He would be Caviar's support and eventually end up bedding her backstage.

"Wouldn't be the first time that's happened, mate," he laughed.

We chatted some more, then it was time for the show. I told Bruce I'd better find John.

I found him down by the side of the stage with the beer.

"Where have you been?" he asked.

"I've been having a drink with Bruce and Spike in the dressing room."
"Bloody great, thanks," he grumbled. "I thought you might be."

"You didn't miss much," I told him. "This whole place is a shit-hole."

Leaning on the edge of the stage was a great place to be for the show.
Two or three times at the end of a song Paul would come over to us
and thrash out his last note, thrusting his long hair over his face and
acknowledge us. People were noticing. It was fun.

We were secretly both hoping that there was going to be some wild,
backstage party.

"Not tonight, Des," said Paul, "There'll be plenty of time for that.
Keith and me have to be at rehearsals in London tomorrow. We're in
another band, The Down and Outs. We're supporting Paul Rogers on
tour. We'll be playing the 02 Arena in Birmingham. You'll have to come
and see us."

That's another story.

We hung around with Bruce and the band as everything was being
packed away. We had called for a taxi. When it came, we said goodbye for
the time being and went back to the Travelodge for a midnight feast and
after-show party.

We were both pretty excited. It had been a great night. We hadn't had a
great deal to drink at the concert and wasted no time opening a bottle of
gin. With ice and cold tonic from the cool box, we talked into the early
hours and didn't get much sleep.

The band were soon to start a European tour and had told us they
would be in Barcelona. As the gin disappeared, our enthusiasm for a trip
to Spain grew strength. By the end of the bottle, we were making travel
plans. Also on the agenda was a trip to see Paul and Keith in the Down
and Outs at the 02 Arena.

Great stuff, that gin!

Daylight found us a little the worse for wear.

A little groggy, we found a place to eat with plenty of coffee. John
dropped me off at home and we agreed to drive everyone at the shop nuts
on Monday.

I think we certainly managed to do that!

• • • • • • • • • •

Pauline phoned me with great news.

"Des, I've done something that could get me into trouble." "What's that?" I asked.

"I've found your manuscript and I've pinched it. It's under my desk."

The receivers had found someone else to take over from Athena and wanted everything returned.

Pauline told me that her company, Fast Print Publishing, was willing to complete the publication of my book and she would be in charge of it. I was over the moon. Over the next few weeks things got moving again.

Pauline and I were having regular Friday afternoon updates and both enjoying watching its progress. John was in contact with the Quireboys and I managed to get a copy of their greatest hits signed by Spike with a little note.

I sent it to Pauline.

She wrote to me:

"Just a little note to say how much I enjoyed the CD. I would say that I got the housework done in record time, but I kept stopping to have a dance! Spike has a great voice by the way.

"Thanks again and I shall look forward to the Quireboys in print - once we've got this book published."

Pauline

Together we designed the cover.

It was a spectacular image of a helicopter exploding in midair. The cover blurb we had inherited from my editor at Athena.

Jack Reec's hurting. Wounded three times by his old enemy, psychopathic killer David Walker, he's got his life back – just - with the help of two fun-loving senoritas. But when Jack starts diving for Brinks Mat gold bars in Malaga Lakes, and is befriended by beautiful sensation Caviar, his Andalusian farm attracts attention from the wrong quarters, including David Walker. The psycho has made a fresh kill and is now branded "The Rabbit Man" and "The Costa Killer."

And then Caviar goes missing...

With the action swinging from Southern Spain to the Black Sea coast of Turkey, the intimacies of kidnapping and revenge are fully explored. It's the Stockholm syndrome, and it's often sexual, sometimes humorous and frequently violent. Desmond McGrath's insight into the minds of killer, captive and pursuer is breathtaking, and in an epic showdown worthy of a Clint Eastwood Western, we see how an armed man who's frightened of nothing can harbour a secret terror - of the beast inside himself.

Pauline sent me the final proofs for my approval. "What do you think, Des?"

I thought it was great. We were going to print!

Fifty

Like the phoenix, I had risen from the ashes. Twelve months ago, my edited manuscript had been lost to the world. A couple of years' hard work down the drain. Now, with Pauline's help, I was back in print. Against all the odds.

Simon Potter was assigned as my Press Officer. He was on the ball. He sent out over twenty copies of the press release to local papers, radio stations and TV channels. He arranged exposure on the Internet and Fast Print's own website.

He sent copies to *The Jersey Evening Post,* BBC Radio Jersey, Channel 103, Island Life and Channel Television. Some time later, I was to visit Jersey for a book-signing at Waterstones and a round of interviews with the above-mentioned.

Mom was over the moon and proud as punch when she saw photos of her son in a two-page interview with the *Sunday Mercury*.

When all the initial excitement ended, as it inevitably does, I decided that I didn't want all the hassle of writing another book. I had done a trilogy and Jack Reec had had a good run. It was bloody hard work and I was nearly sixty-four. I wanted to bow out on a high.

One Last Kill had been placed in "Blockbuster Territory." I had been mentioned in the same breath as Ian Fleming. Graham Young told me that the books would make great films. I should sell the film rights. I phoned him a couple of years later and the first thing he said was, "Des, have you sold the film rights yet?"

It was around this time that Mom had a fall, coming from the shops. I had warned her not to go out in the bad weather, but she was stubborn and obstinate.

After a short spell in hospital, she began to deteriorate slowly. I was having to go in every day to look after her. It was getting too much and I decided to cash in my T-bone steak at Lyons Butchers and retire again.

With all the stress and extra time to kill, my intake of lager increased. Unknown to me, my years of rock-and-roll lifestyle had damaged my liver. I was told to cut down, which I did, but fifty years of accumulated damage does not go away overnight.

I was taking Mom at least once a week to the doctor, the hospital, the foot clinic, or the dentist. Eventually it all caught me up and I had a mini-stroke. It left me slightly disabled, with walking difficulties and no dexterity in my left hand. But it could have been worse.

Mom was 89. She knew she couldn't go on for ever. On one of our many hospital visits, she was told she had cancer.

She took it in her stride and told me that she wanted no one to know. In her broad Irish accent, she told me, "Sure, it's well-known I have to die of something."

She battled bravely to the end. The doctors and Macmillan Nurses wanted me to put her in a home. I was having none of it. If she was going to die, then it would be at home in her own bed.

And so it came to pass. The doctor came in on the afternoon and said that she expected Mom to go in the night. We put her to bed and I got a chair upstairs in the bedroom. Armed with eight cans of Holsten Pils, I sat things out.

About one in the morning I lay on the bed beside her and dozed off. When I woke up about an hour later she was lying staring up at me and holding my hand.

She was dead.

I phoned Barbara immediately and she came straight around. We phoned the nurses and they arrived quickly. They confirmed what I already knew.

I had another can and phoned the funeral directors. They came straight away and removed her from the house. It seemed very empty when she was gone.

The funeral took place quickly and we gave her a great send-off. She would have been proud.

Then came all the hard stuff.

Sensitively trying to sort out somebody's ninety-one years of life. I felt like I was intruding, invading her privacy. But it had to be done and I wanted to do it myself. If she had any secrets and I found them, then it would be only me and not the rest of the world that knew about it.

When it was done, the house went on the market. It was the final wrench. These things take time but, finally, I was to close the door behind me for the very last time.

A huge part of my life was closed in there behind me.

My cousin Sharon, Uncle Bill's daughter, and I had always been close. Her dad was the last of eight brothers and sisters. It was obvious that Bill would likely be the next to go.

We decided to keep in regular contact. Sharon would see my car at the pub and drop in for a drink. Quite often she would surprise me by bringing Bill, too.

On one of our meetings, I confessed that I would love to visit Jersey and try to get in touch with my "family", the Handsfords.

"Why don't you, then?" asked Sharon. "It's only an hour away."

"Barbara won't fly," I told her, "And I can't go on my own. Not with these legs."

"What if I take you?" she volunteered. "I'd love to go to Jersey, especially with you. You know it like the back of your hand. It would be fun."

I thought that was a great idea. "I'll alert the media," I said.

She thought I was joking. I wasn't. I busily set about arranging things. First call was Waterstones in St Helier. Lizzy, the Manager, arranged a book-signing date. Next was Thomas Ogg, at the *Jersey Evening Post*. We set up an interview. Local TV agreed to plug my book and visit on the evening news. BBC Radio Jersey arranged an interview on their breakfast show.

Sharon arranged the travel, hotel and hire car and we were good to go.

We landed in Jersey on a beautiful, sunny afternoon. We collected the hire car and enjoyed the lovely ride along the coast road to St Helier. Our hotel was located in the centre of town to make it easy on my legs.

Putting away our cases in our rooms, we agreed to meet in the bar. It was the last week the hotel was to be open before it closed for the season. It was quiet, and the manager was serving behind the bar. I took a stool and ordered a beer.

"Good evening, sir," he greeted me cheerfully, "Are you here on business or pleasure?" "Both," I told him truthfully. "I was born on the island and have come home to look up some friends and relatives and promote my latest book."

"Your book, sir?" he said, questioningly.

"Yes; I'm doing a book-signing in St Helier and a series of interviews with the local press and media."

"How fascinating," he said, and asked me all about it.

Reluctantly, I told him about the Jack Reec trilogy and the sensitive negotiations for the film rights, as he poured me another pint "on the house". Sensing I was reeling in a big fish, he brushed aside my futile effort to pay for Sharon's large gin and tonic as she slid in by my side.

Introducing my PA, Sharon fell effortlessly into her role and, with grand gestures, told of her tireless efforts to make sure everything was to run smoothly with my radio, TV and newspaper interviews.

With our new celebrity status established beyond doubt, we were escorted, personally, to the table of our preference in the dining room.

Chuckling away to ourselves, we enjoyed an excellent meal with a suitable bottle of red, highly recommended by the maitred'e.

Set for the evening, we set off to find ourselves a local pub for a few pints, intending to have the last one in the hotel.

Next morning, after a super breakfast, we walked to the shops to find Waterstones and Lizzy. She had been expecting us and everything was arranged for Saturday morning. Ten o'clock till two o'clock.

That sorted, we collected the car and, with Sharon driving, I took her to a few good sightseeing spots.

Gory Castle had always been a great favourite of mine. Apart from the castle, there was a nice little harbour with plenty of shops and pubs. After a spot of lunch, Sharon went for a wander and I sat outside a bar in the sun with a pint, watching the world go by.

Tomorrow I was to meet Tom from the *Jersey Evening Post* for an interview. I used my time in the sun finalising the details on my mobile. I was to meet him at ten in the morning at Costa Coffee in St Helier.

When Sharon came back, I filled her in and we drove back to the hotel for dinner and our evening drinks.

In the morning, Sharon came with me into town. I found Tom and did my interview while Sharon went shopping. Everything went well and, by midday, we were on our way to St Ounes on the Five Mile Road. There were German bunkers there that had been turned into a war museum.

Sharon was fascinated with all things historical. It was a very interesting place, that showed how life had been for both the Germans and the islanders under the occupation.

After a couple of hours there, it was lunch in a country pub and a couple of pints while we planned the next interview. BBC Radio Jersey. I was to be at the studios for 7.30 am. Too early for Sharon! Fortunately, it was located only a ten-minute walk from our hotel in the centre of St Helier.

Before dinner, we decided to do a recce and find the place. We found it, no problem. When Sharon was certain I knew exactly where it was, we went to our new favourite restaurant.

Sorry. Did I say restaurant? I meant local café. Finding after the first night that the food in the hotel, for what it was, was horrendously expensive, we had found the *Wagon Wheel*. It was a corner café run by a Polish couple and catered for the local hotel-workers. The food was great and really cheap. It was licensed, so we could have drinks too. It suited us just fine.

I was up at the crack of dawn to make sure that I gave myself plenty of time to get to the studios. I was extremely nervous and apprehensive about it all and must have taken a wrong turn because, soon, I was hopelessly lost and starting to panic. I found myself knocking on the door of a local greengrocer preparing his fruit and veg for the day. He told me through the door that he wasn't open but, eventually, seeing my persistence, he opened the door.

I explained my predicament and he helpfully gave me directions, even walking with me to the corner to point out the direction. I think he could see how flustered and panicked I was and felt sorry for me.

I arrived in a sweat at the studios. The girl on Reception was friendly and calming. I told her how nervous I was and she only laughed. She told me not to worry. I was slotted in for 7.50, just before the 8.00 o'clock news. She told me that the presenter was a really nice guy and would put me totally at ease.

His name was Mike Weiler. He was not a local but was, in fact, from my neck of the woods in England - Stourbridge. His producer came in and introduced herself and gave me a briefing and a cup of tea to steady my nerves.

She also told me what a nice, popular presenter Mike was, and tried to put me at my ease.

The moment came. I was taken into the studio and sat on the other side of a glass screen from the presenter, who was talking to a screen while waving to me. The producer fitted me with large headphones and set the mic up against my mouth.

It was precision stuff. Some sort of hand signals, a countdown, a flurry of activity and then Mike Weiler began.

"Listeners, I'd like to introduce you to Desmond McGrath; he's a local-born author and he's here today to tell us all about his latest book, a thriller called *One Last Kill*. But first, Des, tell us a bit about yourself."

He began asking me questions about my early life in Jersey. At first, I stumbled a bit with my answers, but soon got into my stride and had an extremely good interview.

"Well, Des, it certainly sounds like a very exciting book. I look forward to reading it. And listeners, if you would like to meet Des in person, he will be at Waterstones in St Helier tomorrow from 10.00 till two, signing copies of his book, *One Last Kill*. Thank you for coming on the show, Des, we all wish you luck with your book."

I breathed a sigh of relief as the producer led me away. "That wasn't so bad, was it Des?" she said.

We had a bit of a laugh and a chat as she handed me to another girl, who took me back to Reception.

"You were great," said the girl, "Bet you can't believe what you were worrying about." "Sure," I said.

• • • • • • • • • •

Well, if I thought that was bad, then I was wrong. The worst was yet to come. The book-signing.

I got there at 9.45 am. Lizzy was waiting for me. There was a table with a pile of books on it and a chair, which she told me I should sit on.

I was in my best suit and felt a right prat. Straight away, I decided that I wasn't sitting in the middle of the shop for four hours looking like a right berk.

Sharon certainly wasn't.

"Phone me later and let me know how it's going. I'll meet you in the pub later."

I'm not kidding. This goes down as one of the most embarrassing and awful few hours of my life. A few people approached me and spoke, picked up a copy of the book and put it back. A few people bought a copy, which I signed for them.

The time dragged by. It was only 11.30. Christ. I looked about me. Most customers were in and out, having known exactly what they wanted before they came in.

Others browsed the specialist sections. Nobody wanted to be harassed by the prat in the suit on a boiling-hot day, when all that they wanted was to do their shopping and get to the beach.

Lizzy came to me. "How's it going, Des?"

"I absolutely hate it," I told her. "I've never done a book-signing before and, I swear to God, I'll never do one again.

She looked shocked. "To be fair, Des, it's the quietest Saturday we've had for weeks. It always happens when we have a beautiful day like this. People want to be on the beach or outside a pub.

"Me included," I said. "I'm going for a pint."

There was a pub about a hundred yards up the road. I took off and ordered two pints of Carling. It was busy and I didn't have much time. I wasn't gone long. When I got back it was about 12.30. The shop was almost empty. I stood guard over my books for a while then wandered around the empty shop. Lizzy was on an hour's lunch. I looked at my watch and approached the sales desk.

"Can you say thanks to Lizzy for me," I said to a couple of staff drinking coffee. "Tell her I'll be in touch later, but I'm calling it a day."

With that I went to the pub. It was only 1.00 o'clock. I phoned Sharon and told her the situation. She came for me and we walked back to the hotel. I think we laughed about it all the way. I couldn't explain how awful the whole thing had been.

When we got there, we were greeted by some people who we had been with the night before. They asked how it had gone. I told them truthfully and we had a great laugh. The manager came over to join in the fun. He asked if we were eating in the hotel that night. We told him no, we were going to the *Wagon Wheel*.

Slightly disappointed, he said he would like to buy us a drink before we left, as we were going home the next day. He said he would see us in the bar later. If he wasn't there he would leave instructions to the staff to see that we had drinks on him.

Afraid of offending him, we had our dinner in the café and a few pints in an Irish bar down the town. In plenty of time to accept his hospitality, we went back to the hotel bar where, indeed, instructions had been left for us to have drinks on the house.

The manager didn't appear, but we saw him in Reception after breakfast and thanked him for his splendid hospitality. We took our luggage to the car park and drove to the airport for our flight home.

An hour in the Duty Free and another on the plane, and we were back in Birmingham. We said our goodbyes. Sharon was expecting a lift. I phoned Barbara. She said she would come and collect me.

It was chaos in the car park. I told her that it would be a waste of time. The queue into the airport was a mile long. I had my bus pass on me and the 900 stop was fifty yards away. A bus was due. I caught it and, twenty minutes later, I was stepping off at the bus stop outside my house. Barbara was leaning on the garden fence waiting for me.

She had a dinner cooked for me and beers cold in the fridge. It was nice to be back home.

I had missed her.

I phoned Lizzy at Waterstones first thing Monday to tie up a few loose ends.

"Great news, Des," she laughed. "On the day, you outsold Lee Child and James Patterson." They must have been having a bad day!

Fifty-one

Barbara was excited to hear all about it. She had been listening to my radio interview on her computer in the kitchen. Lol, a mate of mine, was there putting a new radiator in the conservatory.

"Des is on the radio," she said. "Can you stop banging for a minute?" "You what!" he said surprised. "What's he doing on the radio?"

She told him.

"I didn't even know he wrote books," he said incredulously. "He's a bit of a dark horse.

He never said anything."

"Well that's Des," she said, "He doesn't say much."

They both listened, laughing and taking the piss. When she told me this, I knew I was in for a piss-take next time I saw him in the pub.

I wasn't wrong.

Things soon got back to normal. Until, one day, I felt ill. My gammy leg got worse and I was having trouble with the grip in my left hand. I seemed to have lost dexterity. I phoned the doctor. She told me to come straight round. We sat discussing my symptoms, then she said,

"Can you get to the hospital?"

"Of course," I said, "I've got the car, I can drive there."

"I wouldn't recommend that, Des," she said. "I don't think you'll be coming back. I'm phoning A&E at Heartlands. They'll be expecting you. Tell them who you are. You won't have to queue. You'll go straight in. I think you've had a mini-stroke."

I started to panic a bit.

"Can Barbara take you?" she asked. "She's out, but I'll phone her," I said.

"Well, go home and call a taxi if she can't. Get there as soon as you can."

I phoned Barbara and told her what was happening. She said she would get straight back and take me.

Reception at A&E was alerted and I saw a doctor straight away. After a few questions and simple tests, I was told to wait while they found me a bed. They would do an MRI scan first thing the next day.

The scan showed I had a small growth on the brain. It could be new, or something I was born with. They spent all day in a meeting deciding what to do. They couldn't make their minds up whether to operate to remove it or leave well enough alone and keep a close eye on it. The fact that my brother had died from a brain haemorrhage suggested that it could be hereditary.

To my relief, they decided I'd had it from birth and did not want to operate. They sorted out medication for me and let me go home, with outpatient appointments.

I worried about it at first but, as time went by, I started to forget about it. I told my mates that I had a brain abnormality. "Well, that explains a lot, Des," they all agreed.

Charming.

• • • • • • • • •

My tribe of grandchildren was growing. Kimberley, the oldest, was married, with a nice home of her own with Rory.

Jack (Reec) and Tara, my other two with Angela, were teenagers living on the other side of town, doing teenage things.

Ella was the next oldest. She was now about seven. We were very close. She reminded me so much of Kimberley when she was young. We spent lots of time together and she was always interested in what I was doing.

Her Godmother is Emma Willis, the TV personality, who is married to Matt of Busted fame. One day at school she was asked,

"How many famous people do you know, Ella?" "Three," she replied, without hesitation.

"Three," said the teacher. "Who are they, Ella?"

"Auntie Emma, Uncle Matt and Grandad," she replied. "Grandad?" the teacher exclaimed.

"Yes," she said firmly. "He writes books." Bless her.

My garage I've converted into a man cave. It houses my collection of tribal art, glass, paintings and old seventies soda syphons. There's a proper roulette table, with all the chips and gambling paraphernalia. I extended it into the garden to make a beach bar, with all my diving and travelling memorabilia. It once made the *Daily Mail*, who sent a photographer and journalist to do a story on it. I couldn't get rid of him.

Ella and I loved playing roulette. She was turning into a proper little croupier. "Place your bets," she'd shout. "No more bets, please. Zero! Banker takes all."

She didn't like losing and, sometimes, it would get quite heated. She loved it that much that I had a big sign made for the man cave that I hung at the end on the door.

"GRANDELLAS, Bar Restaurant Casino," it said, colourfully, and was decorated with dice and playing cards.

One afternoon we were in the middle of a game. The juke box was playing; I was smoking a cigar and drinking a can. Ella was banker.

"Place your bets."

I spread out my chips on the table. "No more bets."

She spun the wheel. The ball clattered noisily around and, after jumping a few times, landed on red.

I had black.

Squealing excitedly, Ella raked in all the chips.

A voice rose above Whitesnake on the juke box. "Des." It was Natalie.

I turned down the music.

"Hello," I said.

"I'm not sure if this is a good example to be setting my daughter."
"How so?"

"How so? Drinking, gambling, smoking cigars and loud rock music. She's only seven, for God's sake."

Ella and I looked sheepishly at each other.

"I hear what you're saying," I said, putting out my cigar and laying down my can. I turned off the juke box.

"No problem," I said. "Roger that. Over and out."

Natalie shook her head in what looked like bemusement and despair and left us for the safety of the house.

"I think she's gone now, Grandad," said Ella. "Yes," I said, "Let's have a break."

I found our secret stash of chocolate and we sat on the freezer with a drink, arguing whether to continue playing roulette or convert the table to a craps table and roll the dice.

We decided craps. Removing all the roulette accessories and lifting off the top revealed a craps table beneath. We broke out the dice and were soon howling with excitement as we bounced them down the table.

Next to be born was Chloe. She was born with an eye problem that needed a series of operations to correct. Thankfully, they were completely successful and, apart from old photos, you would never know there had been a problem.

The last two born were Jensen and Olivia. They are still very young and I'm only just getting to know them.

So, at the ripe old age of 68, that was how I saw my life. Quality time with my kids and grandkids.

"Tell us one of your stories, Grandad," Ella would say. "Grandad's always got a story."

I would then proceed with a tale of my adventures, while Ella rolled her eyes and looked at Barbara as much as to say, "Here we go again," and carried on doing what she was doing while I rambled on.

Kids!

Angela and John had recently moved into a huge, grand house in Curdworth. It was big enough, easily, for me to get lost in. The indoor swimming pool and jacuzzi area was almost as big as the *Belfry*. With a full-size snooker table and six or eight stables, I called it Southfork.

JR's.

Jack Reec! Get it? Never mind.

They loved to party, and Barbara and myself were frequent visitors. So was my ex-wife, Rita, whom I have no axe to grind with. We are still friends.

She also gets on really well with Barbara. "You did me a great favour, Barb," she told her at one party over a bottle of champagne. "He was a bastard. Always having affairs. Do you know he even had an affair with a married woman for five years and I never knew? Teresa Ward was her name."

Now this really shocked me. How the hell did she know that? Wanting to just let it go away, I never asked. Maybe I will.

It was a bizarre situation. Rita on one side of me and Barbara on the other, talking about me as if I wasn't there. Feeling the need to defend myself, I blurted out, "Well, everybody has to have a hobby!"

Apparently, someone had once said, "That bedroom over his shop has a revolving door!"

I ask you. You don't need that!

All in all, life was pretty good. No pressure.

It was a nice summer and I spent my time playing with my car collection and taking a cruise to the local pub for a pint with my mates.

I still had my beloved Mercedes Sports. I'd always wanted a Jaguar XJ6, and my mate Dereck from the butchers had located one that was fifteen years old but looked like it had just come off the factory line. It had only 50,000 miles on the clock and was owned by some retired gentleman who cherished it until the day he passed over.

I couldn't refuse it at £2,150.

The Jaguar XK8 was also in mint condition for its age and, after receiving a totally unexpected bonus cheque in the post, I purchased it on a whim, much to the disgust of Barbara who protested, in vain, that it was ridiculous to have four cars (she had her own car).

The bullets bounced off me. You are only here once and, anyway, they were worth more now than what I paid for them. I wish the same could be said for her Kia Rio, which had lost more in three years than I'd paid for the two Jags. To further enhance them, I invested in two private plates for them that are also going up in value. DEZ 72 and DEZ 85.

Try telling that to Barbara.

Instead, in a moment of mischief, I casually told her, "I'm thinking of getting another Jensen; they're beginning to soar in value, I was reading in the *Classic Car Magazine*."

As my grandchildren might read this book someday, I cannot print her response, as I wouldn't like the image they have of her to be shattered.

It was February 2nd 2017 when I had an urgent message from my liver. "I've had enough," it told me.

I was sitting up in bed watching TV with Barbara. I had a pint of orange pop that I was just finishing when I felt a little queasy. I coughed lightly and tasted something tinny in my mouth. I spat it into the empty glass by the bed. It was thick, red blood.

I felt a vomit coming on and snatched up the glass to catch it. I was left holding three quarters of a pint of warm, sticky blood.

I was frightened. Barbara called an ambulance. They categorised me an emergency and came at a fast response.

A&E was a bit shambolic. It was crazily busy. I was sent to an assessment ward that was equally shambolic. One doctor was talking about sending me home. Another doctor argued that "He's obviously bleeding from somewhere. We can't just send him home!"

Barbara agreed and, in no uncertain terms, told them, "He's going nowhere." They found me a bed.

However, in the early morning, about six, I was talking to the ward sister, who told me it was quite likely that I was going home. That's what she had heard.

I phoned Barbara and told her what had been said.

She was furious and immediately got on the phone to whoever it was at the top. They tried fobbing her off, but she was having none of it and demanded to meet the hospital administrators, telling then that if anything happened to her husband they would be up to their ears in lawsuits and publicity.

They didn't seem keen on that and agreed to see her at the hospital. Coincidentally, about

9.00 o'clock, a team of doctors arrived at my bedside, Barbara not far behind.

A really lovely man was in charge, Dr Wilson, a consultant. We shook hands, exchanged pleasantries. Barbara politely but firmly explained that there was no question of me being turned out of the hospital until I had been treated and recovered.

Dr Wilson couldn't have been more apologetic. "I don't know who gave you that information, Des," he told me, "But whoever it was had no right. I am the man in charge around here and I can assure you that you will not leave this hospital until everything possible has been done for you."

With that he declared me Nil by Mouth and said that I would be collected later in the morning and taken down for an endoscopy. Shaking my hand, he told me he was looking forward to seeing me later and that he would be doing the procedure himself.

"Thank you, doctor," I told him as he left.

I sat with Barbara as she told me how many noses she had put out of joint that morning.

Politely but firmly, she had left them with no illusions about where they stood if they didn't do their job properly.

Take note. Don't be fobbed off. I've seen it time after time. It's the survival of the fittest.

You have to fight your corner. The quiet ones get pushed aside. Don't be rude or shout. Just firmly let them know that you won't settle for anything less than what you're entitled to.

It shouldn't be like that.

But it is.

Remember. They are like you. They also want a quiet life.

Later that morning, a cheerful West Indian porter came for me to take me for the camera.

In a wheelchair, he pushed me for miles through the hospital to the Endoscopy Unit. Inside, I was greeted cheerfully by a group of mixed-race nurses. Examining my ID tag, one said, "And how are you today, Des?"

"Marvellous," I replied, "Never felt better in my life."

Reminding me of the funny policeman in *"Death in Paradise"*, Duane, the porter, stepped back and said, "Desmond, you just told the biggest porky I've heard all day in dis 'ospital. Shame on you."

The camera down the throat was horrible. I had ruptured veins in the oesophagus and needed bands put around them. It seemed that a playboy lifestyle and drinking from the age of fourteen had finally worn down my liver. Not being able to cope, it had pushed my blood into the oesophagus, which couldn't take it and burst, causing massive internal bleeding. Ten years ago, there would have been only a 20% survival rate. As it was, now, it was 50%. I was lucky.

They told me I would be foolish to drink again. Next time would be fatal. No doubt about it.

It wasn't what I wanted to hear. But I was fortunate to be able to hear it. It was going to mean a huge adjustment in my life.

However, while I was in hospital I received all the incentive anyone could need to give up the beer. Chloe, my three-year-old granddaughter, was at nursery. The nursery was concerned about her. She was quiet and withdrawn and crying in a corner. They asked her what was the matter. She blurted out, "My grandad's in hospital."

When I was told, I filled up and couldn't help myself from crying. The thought of it upset me beyond words. Natalie, Ella's mom, also told me, "You know, Des, if anything happened to you, Ella would be devastated."

That made up my mind. I have never touched a drop since.

• • • • • • • • •

I had more cameras down and more veins tied off. It was a shitty time. I'd been in hospital for two weeks. One morning, after breakfast, I was dozing in bed.

I opened my eyes and a tall man in black, with a grey beard and holding a Bible, was standing over me. I looked up at him.

He said, "God is looking down on you, Des."

"Well, tell him I'm very grateful," I said. "I don't mind him looking down on me, I just hope he's not waiting for me."

Colin, my mate in the opposite bed, had visitors, his wife and daughter. They were pissing themselves laughing.

The priest was Catholic, as I am, and I told him my priest was Father John from St Thomas More in Sheldon. He knew him and we chatted a while. He made the sign of the cross and blessed me.

"If you're still here next week I'll see you again, Des."

"I hope not. I want to be out of here by then."

I was. Thank God.

Fifty-two

It was great to be home. Honey the cat had missed me and was driving me nuts. She wanted to sit on my shoulder in my chair and nuzzle her face against mine.

I was weak, but alive and on the mend. I had to see my doctor once a month to check me over. She has looked after me and my family for nearly forty years. Somehow, Dr Susan never seemed to age. She still looks like a fresh young doctor and full of enthusiasm. Leaning back in her chair, she appraised me with a grin.

"I thought we'd lost you this time, Des. How are you feeling?" "Marvellous," I said. "Never felt better in my whole life." "Hmm," she replied, sceptically.

After fifteen minutes of reviewing everything, she gave me an appointment for four weeks' time. I thanked her and got up to leave.

"You've got nine lives, Des," she said, leaning back in her chair, "And you've used up most of them."

"Thanks, doc."

It was early spring and I was spending my time reading, sunbathing when I could and going to the pub for an hour to see my pals. I was drinking pints of orange squash and soda water. Nobody took much notice or took the piss. A few were thinking, "There but for the grace of God."

There were Kimberley's, Tara's and my birthdays in August. Angela was throwing a party at Southfork. For such occasions I had found a great beer, St Peters Without, that tasted like real ale but was non-alcoholic. I took some with me. It avoided all the questions about why I wasn't drinking.

I was talking to Kimberley and Rory in the garden when Kimberley said, "I was watching a programme the other night about Barry Gibb. He reminded me very much of you. He looks a bit like you and he, too, lost both his brothers. I just thought that you've led an interesting life. Have you ever thought about writing your life story? I'd love to read it. It would be something to leave behind for the rest of the grandchildren too. They would know the real you. Not the old man that they see today. Not that I'm saying you look old, but you will be seventy next year."

I had been thinking about it for years but had always decided against it. I didn't want to sound pretentious. However, the context Kimberley had put it in, it made a lot of sense. Most people die and all they leave behind is a box of old photographs. The odd person will recall an old memory and maybe a forgotten story about you.

The truth is, though, that your whole life is over and forgotten. Nobody really knows the real you.

Maybe nobody cares.

But Kimberley cares, and that was good enough for me.

"I'll give it a go," I told her. "I've never written anything like an autobiography before. I don't know the format. What I think is interesting, other people might find boring."

"Knowing you, Grandad, I'm sure you'll find a way. You were always pretty good at improvising."

• • • • • • • • • •

I started by researching my whole life. I dug through letters, photos, newspaper cuttings, diving memorabilia, long-forgotten letters, and found myself remembering things about my life that I had long forgotten.

I started writing. It was hard work. I was riddled with self-doubt. Had I got it right?

Should I mention this and leave out that? I was muddling along, clueless. It wasn't like a Jack Reec novel, making it up as you went along.

I'm conscious about other people's feelings. Sometimes I got stuck in a quandary and had to pause for a few days, to think it out.

It was Rory's mom's 60th birthday party at a club in Erdington. I'd been working for about three months.

"How's the book going?" Kimberley said. I told her.

"How far have you got?" Rory asked.

"I don't really know," I answered honestly.

"How old are you where you've got to?" he said.

"Twenty-four," I answered and, for the first time, realised the enormity of the task. I was only twenty-four and the story had hardly started.

I secretly wanted the book finished and published for my seventieth birthday. At this rate, there was no chance. I decided to try and write something every day. I started stopping in all day, not going to the shops or pub.

It worked. You get a lot more done in the morning than you do in the afternoon. It was getting written, but whether it was any good or not was anyone's guess. I'm sure you lot will tell me.

I wrote all over Christmas and the New Year, and it is now February.

I'm just about done.

No pressure, Kimberley, but it was your idea. If I've fallen flat on my face, it's all your fault.

Happy days.

Epilogue

Pauline at Fast Print was excited about the book launch. She suggested a pre-launch publicity tour to promote the biography and re-launch of the Jack Reec Trilogy, on the back of it.

She had arranged a gruelling schedule of book-signings across Europe, North America, Australia and New Zealand.

She was also in frantic negotiations for the Jack Reec film rights and merchandising. There was a bidding war going on in Hollywood, the most notable producer being Steven Spielberg.

We had conquered Europe and the Eastern Seaboard of America and had flown over to the West Coast.

Sharon had taken time out to accompany me as my PA and Press Secretary. We were having a good time, staying in the best hotels and living a celebrity lifestyle.

I was in my suite at the Beverly Hills *Hilton*, being interviewed by half a dozen beautiful young women from various television networks.

Sharon was taking notes.

I was smoking a White Owl cigar and drinking iced champagne. The phone rang. Sharon picked up.

"It's Steven Spielberg."

"Tell him I'm busy."

"He says it's important. He's willing to go two more million."

"Ask him to ring back."

"He's got to speak to you."

"OK, give me the phone."

Sharon passed it over.

"This better be good, Spielberg. I'm in the middle of something."

I started to feel groggy.

I was being shaken.

"Wake up, Des. Wake up, Des."

It was Barbara.

"You were having a bad dream."

"No I wasn't. It was a great dream."

To be continued....